Cohomological Invariants in Galois Cohomology

University

LECTURE

Series

Volume 28

Cohomological Invariants in Galois Cohomology

Skip Garibaldi
Alexander Merkurjev
Jean-Pierre Serre

American Mathematical Society
Providence, Rhode Island

2000 *Mathematics Subject Classification.* Primary 12G05, 11E72.

For additional information and updates on this book, visit
www.ams.org/bookpages/ulect-28

Library of Congress Cataloging-in-Publication Data

Garibaldi, Skip, 1972–
Cohomological invariants in Galois cohomology / Skip Garibaldi, Alexander Merkurjev, Jean-Pierre Serre.
p. cm. — (University lecture series ; v. 28)
Includes bibliographical references and indexes.
ISBN 0-8218-3287-5 (alk. paper)
1. Invariants. 2. Galois cohomology. 3. Linear algebraic groups. 4. Number theory. I. Merkurjev, Alexander, 1955– II. Serre, Jean-Pierre. III. Title. IV. University lecture series (Providence, R.I.) ; 28.

QA169.G37 2003
514′.23—dc21 2003048151

∞ The paper used in this book is acid-free and falls within the guidelines
established to ensure permanence and durability.
Visit the AMS home page at http://www.ams.org/

10 9 8 7 6 5 4 3 2 1 08 07 06 05 04 03

Contents

Preface

The invariants we are going to discuss are the analogues for Galois cohomology of the characteristic classes of topology (Chern classes, Stiefel-Whitney classes, etc.), where the topological spaces are replaced by the scheme Spec(k) for k a field. Historically, one of the first examples of cohomological invariants of the type considered here was the Hasse-Witt invariant of quadratic forms, defined in Witt's seminal paper [Wi 37]. Later, more invariants of quadratic forms were defined, for example Stiefel-Whitney classes and the Arason invariant. The first work in this volume classifies invariants of quadratic forms and étale algebras with values in Galois cohomology modulo 2 or in the Witt ring. The invariants of some other algebraic structures are also determined. A principal tool is the notion of versal torsor, which is an analogue of the universal bundle in topology.

For G a simple simply connected algebraic group, Rost proved the existence of a canonical and nontrivial invariant of G-torsors with values in dimension 3 Galois cohomology. The Arason invariant of quadratic forms is a special case of Rost's invariant, see p. 107. The second work in this volume gives detailed proofs of the existence and basic properties of the Rost invariant. This is the first time that most of this material appears in print.

The two parts of this book are really separate works. The first began as lecture notes that Serre gave at UCLA in January 2001 as part of the Gill Distinguished Lecture Series. They have been expanded and now include many results that were not mentioned during the lectures, especially on étale algebras of low rank and their applications to Noether's problem on invariants. The second part was written independently by Merkurjev (with one section by Garibaldi). As a consequence, the definitions and notations used in the two parts differ somewhat. For the reader's convenience, we have included common indexes and a common bibliography at the end of the book.

The second author is grateful to the NSF for partial support (grant DMS 0098111).

This book is dedicated to Markus Rost. The influence of his ideas can be seen throughout.

Skip Garibaldi	Alexander Merkurjev	Jean-Pierre Serre
Atlanta, Georgia	Los Angeles, California	Paris
USA	USA	France

Cohomological invariants, Witt invariants, and trace forms

By Jean-Pierre Serre

Notes by Skip Garibaldi

Contents

Introduction

As was already said in the preface, these notes explain how to determine the cohomological invariants for various kinds of objects, especially quadratic forms and étale algebras. We have also included "Witt invariants", where the Witt ring replaces cohomology.

The content of the different chapters is as follows:

Chapter I introduces the different functors which are used later, such as Quad_n, Et_n, and $H(*, C)$.

The next two chapters contain basic material on Galois cohomology, for instance the notion of residue (in the local case, cf. Chapter II) and the elementary properties (residue formula) of the cohomology of purely transcendental extensions (Chapter III). There is little here which is new, but since it was not possible to give precise references — especially about sign conventions — we have included essentially complete proofs.

Chapter IV gives several specialization properties of cohomological invariants (the main one being an unpublished result of Rost).

Chapter V contains the basic properties of the restriction and corestriction of invariants, in perfect analogy with the case of group cohomology, cf. Cartan-Eilenberg [CE 56, Chap. XII].

Chapter VI applies the techniques of Chapters III and IV to the determination of the invariants for quadratic forms, octonions, Albert algebras, etc. In each case, a very helpful fact is that there is a "versal" object whose field of definition is purely transcendental over the ground field.

The last three chapters are mainly concerned with the case of étale algebras. This case was the starting point of the theory: see the letter of Serre to Garibaldi appended at the end of these notes. The main results were given in a series of lectures at the Collège de France: see the résumé [Se 94] and also E. Bayer's report [Bay 94]. Up to now, the proofs had not been published. Among the results are:

- the fact that all cohomological and Witt invariants of étale algebras are derived from just one invariant, the trace form;
- an explicit description of all possible trace forms of rank ≤ 7;
- a proof that Noether's problem (on the rationality of the field of invariants) has a negative answer over $\mathbf{Q}$ for groups such as $\mathbf{SL}(2, \mathbf{F}_7)$, $2.A_6$, or the quaternion group Q_{16} of order 16.

This portion of the book ends with letters from Rost, Serre, and Totaro.

A few complements have been given throughout the text in the form of "exercises". We hope the reader will find them interesting (or challenging) enough.

CHAPTER I

The notion of "invariant"

1. Definitions

We fix a ground field k_0, and consider the category $\text{Fields}_{/k_0}$ of field extensions k of k_0 and two functors

$$A\colon \text{Fields}_{/k_0} \longrightarrow \text{Sets}$$

and

$$H\colon \text{Fields}_{/k_0} \longrightarrow \text{Abelian Groups}$$

In the applications, A will be, for instance, the set of isomorphism classes of étale k-algebras of a given rank, and H will be the mod 2 Galois cohomology of the field k. (For more examples, see Sections 2 and 4.)

DEFINITION 1.1. An *H-invariant of A* is a morphism of functors

$$a\colon A \to H.$$

Here, we view H as a functor with values in Sets. Hence, giving $a: A \to H$ means giving, for every $k \in \text{Fields}_{/k_0}$, a map $a_k\colon E \mapsto a(E)$ of $A(k)$ into $H(k)$ such that, if $\phi\colon k \to k'$ is a morphism in $\text{Fields}_{/k_0}$, the diagram

$$\begin{array}{ccc} A(k) & \xrightarrow{a_k} & H(k) \\ \downarrow & & \downarrow \\ A(k') & \xrightarrow{a_{k'}} & H(k') \end{array}$$

is commutative.

Our aim will be to determine explicitly in some cases (e.g., the one mentioned above) the group $\text{Inv}(A,H)$ of all such invariants. If it is necessary to emphasize the role of the base field k_0, we will write $\text{Inv}_{k_0}(A,H)$.

REMARK 1.2. It is well-known that considering the "set" of all field extensions of k_0 may lead to logical difficulties. But in all cases we will consider, the functor A commutes with direct limits, so that an invariant a is determined by its restriction to the category of finitely generated extensions of k_0 (which one can further assume to be contained in a large enough algebraically closed extension of k_0). Moreover, in most cases, there is a finitely generated extension k_1 of k_0 and an element E in $A(k_1)$ such that the knowledge of a on E determines a, i.e., the map $\text{Inv}(A,H) \to H(k_1)$ given by $a \mapsto a(E)$ is injective, see Section 12.3.

2. Examples of functors A

In the examples below, we only give the value of the functor A for an extension k of k_0. The arrow $A(k) \to A(k')$ associated with a k_0-embedding $k \to k'$ is obvious ("extension of scalars").

EXAMPLE 2.1 (Étale algebras [Bou 59, V.6]). Let n be a positive integer. We define the functor A by

$$k \mapsto \mathrm{Et}_n(k) = \{\text{isomorphism classes of étale algebras over } k \text{ of rank } n\}.$$

Recall that an *étale k-algebra* E of rank n is a commutative k-algebra such that $E \cong \prod_i k_i$ for k_i finite separable field extensions of k and $\sum_i [k_i : k] = n$. An equivalent definition is that E is commutative, $[E : k] = n$, and the k-bilinear form defined on E by $(x, y) \mapsto \mathrm{Tr}_E(xy)$ is nondegenerate.

The two most important cases are the one where E is a field, and the one where E is *split*, i.e., isomorphic to $k \times \cdots \times k$.

EXAMPLE 2.2 (G-Galois algebras [BS 94, 1.3], [KMRT 98, 18.15]). Let G be a finite group. A *G-Galois algebra* over k is an étale k-algebra E with an action of G such that G acts simply transitively on $\mathrm{Hom}(E, k_s)$, where k_s is a separable closure of k. Such an algebra is determined up to isomorphism by a continuous homomorphism $\varphi\colon \mathrm{Gal}(k_s/k) \to G$, and two such homomorphisms give isomorphic G-Galois algebras if and only if they are conjugate by an element of G. We set:

$$G\text{–}\mathrm{Gal}(k) = \{\text{isomorphism classes of } G\text{-Galois algebras over } k\}.$$

EXAMPLE 2.3 (Central simple algebras). For $n \geq 1$ we set

$$\mathrm{Cent.\,Simpl}_n(k) = \{\text{isomorphism classes of central simple } k\text{-algebras of rank } n^2\}.$$

In the next four examples, we assume $\mathrm{char}(k_0) \neq 2$.

EXAMPLE 2.4 (Quadratic forms). For $n \geq 1$, we set

$$\mathrm{Quad}_n(k) = \left\{\begin{matrix}\text{isomorphism classes of nondegenerate quadratic forms}\\ \text{over } k \text{ of rank } n\end{matrix}\right\}.$$

We write $\langle \alpha_1, \ldots, \alpha_n \rangle$ for the diagonal quadratic form defined by the α's.

EXAMPLE 2.5 (Quadratic forms with specified discriminant). For a fixed $\delta \in k_0^*/k_0^{*2}$ and $n \geq 1$, set

$$\mathrm{Quad}_{n,\delta}(k) = \left\{\begin{matrix}\text{isomorphism classes of nondegenerate quadratic forms}\\ \text{over } k \text{ of rank } n \text{ and discriminant } \delta\end{matrix}\right\}.$$

Recall that the discriminant[a] $d(q)$ of a quadratic form $q = \langle \alpha_1, \ldots, \alpha_n \rangle$ is the product $\alpha_1 \cdots \alpha_n$ viewed as an element of k^*/k^{*2}.

EXAMPLE 2.6 (Hermitian forms). Let $k_1 = k_0(\sqrt{\delta})$ be a quadratic extension of k_0. The functor $\mathrm{Herm}_{n,k_1/k_0}$ (abbreviated Herm_n) is defined as follows: If k is an extension of k_0, $\mathrm{Herm}_n(k)$ is the set of isomorphism classes of nondegenerate hermitian forms of rank n over the quadratic k-algebra $k \otimes_{k_0} k_1 = k[X]/(X^2 - \delta)$, cf. e.g. [Knus 91, §I.3]. When δ is not a square in k, we have $k \otimes_{k_0} k_1 = k(\sqrt{\delta})$; this is the usual setting of hermitian theory. When δ is a square in k, we have $k \otimes_{k_0} k_1 = k \times k$, and one finds that all the hermitian forms of rank n are isomorphic: the set $\mathrm{Herm}_n(k)$ has only one element.

For $\alpha_1, \ldots, \alpha_n \in k^*$, we write $\langle \alpha_1, \ldots, \alpha_n \rangle_H$ for the corresponding diagonal hermitian form.

---

[a]In [KMRT 98], this is called the "determinant" of q, and the (signed) discriminant is defined to be $(-1)^{n(n-1)/2} d(q)$.

EXAMPLE 2.7 (Pfister forms). For $n \geq 0$,

$$\mathrm{Pfister}_n(k) = \{\text{isomorphism classes of } n\text{-Pfister forms over } k\}.$$

Recall that a 1-Pfister form is a binary quadratic form of type $\langle 1, \alpha \rangle$. An n-Pfister form is a tensor product of n 1-Pfister forms. Its rank is 2^n.

EXAMPLE 2.8 (Octonion algebras).

$$\mathrm{Oct}(k) = \{\text{isomorphism classes of octonion } k\text{-algebras}\}.$$

Recall that an *octonion algebra* O over k is an 8-dimensional nonassociative k-algebra which is a composition algebra [SV 00, 1.6].

Note that, if $\mathrm{char}(k_0) \neq 2$, the functor Oct is isomorphic to the functor $\mathrm{Pfister}_3$ by

$$\text{octonion algebra} \mapsto \text{norm form},$$

see [KMRT 98, 33.18, 33.19] or [SV 00, 1.7]. This extends to characteristic 2 with a suitable definition of Pfister forms, see e.g. [Se 95, §10.2].

EXAMPLE 2.9 (Albert algebras).

$$\mathrm{Alb}(k) = \{\text{isomorphism classes of Albert } k\text{-algebras}\}.$$

Recall that an *Albert k-algebra* is a 27-dimensional exceptional Jordan algebra, cf. [SV 00, Ch. 5], [PR 94], or [KMRT 98, Ch. IX].

3. Torsor interpretation of the above examples

3.1. TORSORS. All of the examples in the previous section (except for Example 2.7 for $n > 3$) have an interpretation in terms of the functor $\mathrm{Torsors}_G$ of (isomorphism classes of) *G-torsors*, where G is a smooth linear algebraic group over k_0.[b] Note that $\mathrm{Torsors}_G(k)$ is the same as the first *nonabelian Galois cohomology set* of G, usually denoted by $H^1(k, G)$, see e.g. [Se 65, Ch. III] or [KMRT 98, Ch. VII]. For instance, in Example 2.4 (quadratic forms of rank n) one selects a *base point* q_0 in $\mathrm{Quad}_n(k_0)$, e.g., the form $\langle 1, 1, \ldots, 1 \rangle$. One defines G to be the *orthogonal group* of q_0, i.e., $G = \mathrm{Aut}(q_0) = \mathbf{O}_n$; this is a smooth linear algebraic group over k_0. If q is an arbitrary element of $\mathrm{Quad}_n(k)$, the variety $P_q = \mathrm{Isom}(q_0, q)$ is a G-torsor over k. The map $q \mapsto P_q$ is a bijection of $\mathrm{Quad}_n(k)$ with $\mathrm{Torsors}_G(k)$, i.e., with $H^1(k, G)$, cf. [Se 65, Ch. III].

In all these cases, when one has selected a base point q_0 for the functor A (so that A can be identified with the functor $k \mapsto H^1(k, G)$, with $G = \mathrm{Aut}(q_0)$), this base point corresponds to the element "1" of $H^1(k_0, G)$.

EXAMPLE 3.2. For $A = \mathrm{Et}_n$ as in 2.1, we may take the split algebra

$$E^{\mathrm{split}} = k_0 \times \cdots \times k_0$$

as a base point. Then $G = \mathrm{Aut}(E^{\mathrm{split}}) = S_n$ (permutations of $\{1, 2, \ldots, n\}$), viewed as an algebraic group of dimension zero.

Let $\overline{k}$ be an algebraic closure of k, and let k_s be the separable closure of k in $\overline{k}$; put[c] $\Gamma_k = \mathrm{Gal}(k_s/k) = \mathrm{Aut}(\overline{k}/k)$. The set $H^1(k, S_n)$ may be identified with the set $\mathrm{Hom}_{\mathrm{conj}}(\Gamma_k, S_n)$ of continuous homomorphisms $\Gamma_k \to S_n$ up to conjugation.

---

[b]If it is useful to mention k_0 explicitly, we use the heavier notation $\mathrm{Torsors}_{G/k_0}$.

[c]The symbols $\overline{k}$, k_s, and Γ_k will be used without further explanation throughout this part of the book.

The correspondence between classes of n-étale algebras over k and elements of $\mathrm{Hom}_{\mathrm{conj}}(\Gamma_k, S_n)$ is just Galois theory, and for such an algebra E we write φ_E for the corresponding homomorphism $\Gamma_k \to S_n$, see [BS 94, 1.3.1]; note that it is only defined up to conjugation. The orbits of $\mathrm{im}(\varphi_E)$ in $\{1, \ldots, n\}$ correspond to the factors of E which are fields. For instance, E is a field if and only if $\mathrm{im}(\varphi_E)$ is a transitive subgroup of S_n, and E is split if and only if $\varphi_E = 1$.

Here is a list of groups G corresponding to other functors A from Section 2:

2.2: The spectrum of a G-Galois algebra is a G-torsor and conversely, see [KMRT 98, p. 290]. Hence a G-Galois algebra over k corresponds to an element of $H^1(k, G)$.

Note that we have two interpretations of $H^1(k, S_n)$: as classes of rank n étale algebras and as classes of S_n-Galois algebras. One passes from the first interpretation to the second as follows: if L is an S_n-Galois algebra, the fixed points of S_{n-1} on L make up an étale algebra E of rank n. Conversely, if E is étale of rank n, one may construct the corresponding L as a suitable quotient of $E \otimes \cdots \otimes E$ (n-fold tensor product); in geometric language, one removes the "fat diagonal" from the product of n copies of $\mathrm{Spec}\, E$.

2.3: Base point the matrix algebra M_n; then $G = \mathrm{Aut}(M_n) = \mathbf{PGL}_n$.

2.4: $G = \mathbf{O}_n = \mathbf{O}(q_0)$, see above.

2.5: $G = \mathbf{SO}(q_0)$, where $q_0 = \langle 1, \ldots, 1, \delta \rangle$.

The exact sequence $1 \to \mathbf{SO}(q_0) \to \mathbf{O}(q_0) \to \{\pm 1\} \to 1$ gives an embedding of $H^1(k, \mathbf{SO}(q_0))$ in $H^1(k, \mathbf{O}(q_0))$ which can be used to identify $H^1(k, \mathbf{SO}(q_0))$ with $\mathrm{Quad}_{n,\delta}$, see [KMRT 98, p. 407].

2.6: $G = \mathbf{U}_n$, where $\mathbf{U}_n$ is the unitary group of rank n relative to k_1/k_0 (and relative to the hermitian form $\langle 1, \ldots, 1 \rangle_H$, say). The fact that $\mathrm{Herm}_n(k)$ has only one element if $\delta \in k^{*2}$ corresponds to the fact that $\mathbf{U}_{n/k}$ is then isomorphic to $\mathbf{GL}_n$, hence has a trivial H^1.

2.8: Base point the split octonions [SV 00, p. 19]; $G =$ split group of type G_2.

2.9: Base point the split Albert algebra [KMRT 98, p. 517]; $G =$ split group of type F_4.

REMARK 3.3. For $n = 1, 2, 3$, the n-Pfister forms have the alternative descriptions:

$$\mathrm{Pfister}_1 \cong \mathrm{Quad}_1 \cong \mathrm{Et}_2 \cong H^1(*, \{\pm 1\}),$$

$$\mathrm{Pfister}_2 \cong \mathrm{Quad}_{3,1} \cong \mathrm{Cent.\ Simpl}_2 \cong H^1(*, \mathbf{PGL}_2) \cong H^1(*, \mathbf{SO}_3),$$

$$\mathrm{Pfister}_3 \cong \mathrm{Oct} \cong H^1(*, G_2).$$

However, for $n \geq 4$, one can show that there is no algebraic group G over k_0 such that the functor $\mathrm{Pfister}_n$ is isomorphic to the functor $k \mapsto H^1(k, G)$.

NOTATION 3.4. When the functor A is the functor $k \mapsto H^1(k, G)$, we write $\mathrm{Inv}(G, H)$ for $\mathrm{Inv}(A, H)$.

4. Examples of functors H

4.1. ABELIAN GALOIS COHOMOLOGY. Let C be a discrete Γ_{k_0}-module. For any extension k/k_0, we have a natural map $\Gamma_k \to \Gamma_{k_0}$ (defined up to inner conjugation), so that the cohomology groups $H^i(\Gamma_k, C)$, $i = 0, 1, \ldots$ make sense, cf. [Se 65,

II.1.1]; they are denoted by $H^i(k,C)$, and their direct sum is written $H(k,C)$. These groups are functorial in k, i.e., they define functors

$$\text{Fields}_{/k_0} \to \text{Abelian Groups}.$$

These will be the functors "H" that we will consider most of the time, and for such H we write $\text{Inv}^i(A,C)$ or $\text{Inv}(A,C)$, instead of $\text{Inv}(A,H)$; similarly, if $A = \text{Torsors}_G$, where G is as in 3.1, we use the notation $\text{Inv}^i(G,C)$ or $\text{Inv}(G,C)$. The elements of $\text{Inv}(A,C)$ and $\text{Inv}(G,C)$ are called *cohomological invariants.* When it is necessary to mention k_0 explicitly, we write k_0 as a subscript, as in $\text{Inv}_{k_0}(G,C)$.[d] Most of the time we will assume that C is finite and of order not divisible by the characteristic, the principal example being $C = \mathbf{Z}/2\mathbf{Z}$.

EXAMPLE 4.2. Let G be a finite group and let C be a G-module with trivial action. Write $H(G,C)$ for the direct sum of the cohomology groups $H^i(G,C)$. An element x of $H^1(k,G)$ is represented by a continuous homomorphism $\varphi_x : \Gamma_k \to G$; such homomorphisms represent the same class x if and only if they are conjugate. The map φ_x gives a map $\varphi_x^* : H(G,C) \to H(k,C)$ which depends only on x (but not on the choice of φ_x). If u is a given element of $H(G,C)$, define $a_u(x)$ as $\varphi_x^*(u)$. One gets in this way a cohomological invariant a_u which belongs to $\text{Inv}(G,C)$ in the notation of 4.1 above.

For a concrete example of this construction when $G = S_n$, see 25.1.

Note that even when u is not 0, the corresponding invariant a_u may be zero for all possible choices of k_0. Such a class u is called *negligible*, see §26.

4.3. Another example of a functor H is $H(k) = W(k)$, the Witt ring of nondegenerate quadratic forms on k modulo hyperbolic forms, see Chapter VIII.

DEFINITION 4.4. There is a natural embedding $H(k_0) \to \text{Inv}_{k_0}(A,H)$; namely, if h belongs to $H(k_0)$, we define the invariant a_h by setting

$$a_h(x) = \text{image of } h \text{ in } H(k)$$

for every $x \in A(k)$. Such an invariant is called *constant.*

4.5. Suppose we have fixed a base point for A. For example, if $A(*) = H^1(*,G)$ as in §3, the base point is the unit element "1". We say that a is *normalized* (relative to the choice of base point for A) if a vanishes on the base point. Every invariant can be written in a unique way as (constant) + (normalized).

5. Versal objects

Let G be a smooth affine algebraic group scheme defined over k_0.

DEFINITION 5.1. A *versal G-torsor* is a G-torsor P over a finitely generated extension K of k_0 such that there exists a smooth, irreducible variety X over k_0 with function field K and a G-torsor $Q \to X$, with base X, with the following two properties:

(1) The fiber of Q at the generic point of X is P.

[d] Our general policy is to use shortened notation as much as possible.

(2) For every extension k of k_0 with k infinite, every G-torsor T over k, and every nonempty open subvariety U of X, there exists $x \in U(k)$ whose fiber Q_x is isomorphic to T (i.e., the set of $x \in X(k)$ with $Q_x \cong T$ is dense in X).

EXERCISE 5.2. Show: If (P, K) is a versal G-torsor as above, the field K is a regular extension of k_0, i.e., the k_0-variety X is absolutely irreducible.

5.3. VERSAL G-TORSORS EXIST. Choose an embedding of G in $\mathbf{GL}_N$ for some N. Let X be the homogeneous space $\mathbf{GL}_N/G$. Put $Q = \mathbf{GL}_N$, viewed as a G-torsor over X. *The fiber P of Q at the generic point of X is versal.* Indeed, if T is as in 5.1.2, it follows from [Se 65, I.5.4, Cor. 1], combined with the fact that every $\mathbf{GL}_N$-torsor is trivial, that T is isomorphic to Q_x for some $x \in X(k)$, and that such an x is unique up to the action of $\mathbf{GL}_N(k)$ on $X(k)$. Let O_T be the corresponding orbit of $\mathbf{GL}_N(k)$. Since k is infinite, $\mathbf{GL}_N(k)$ is dense in $\mathbf{GL}_N$ for the Zariski topology, and O_T is dense in X. Hence P is versal.

Note that this gives more. It shows that the functor

$$k \mapsto H^1(k, G)$$

is isomorphic to the functor $k \mapsto X(k)/\mathbf{GL}_N(k)$.

EXAMPLE 5.4. As in 5.3, choose an embedding of G in $\mathbf{GL}_N$ for some N over k_0; write V for the affine N-space on which $\mathbf{GL}_N$ (and hence G) acts.

We also make the following assumption:

(F) *The scheme-theoretic stabilizer in G of the generic point in V is trivial.*

Property (F) implies — cf. [Th 86, Prop. 4.7] — that there is an open dense subset V' of V which is G-stable, and is a G-torsor. Let $X = V'/G$, and let K be the function field of X over k_0; by taking the generic fiber of $V' \to X$, we get a G-torsor P over K. *The torsor P is versal.* Indeed, because of the definition of versal, we may assume that k_0 is infinite, hence V' has a rational point. One can use this point to define a G-equivariant map $\mathbf{GL}_N \to V$, hence a G-equivariant dominant map $P \to V'$, where P is open dense and G-stable in $\mathbf{GL}_N$. The fact that $\mathbf{GL}_N$ has the versal property 5.1.2 implies the same property for V'.

(Note that the dimension of X in 5.4 is $N - \dim G$ which is typically much lower than $\dim(\mathbf{GL}_N/G)$ in 5.3.)

EXAMPLE 5.5. If G is *finite*, then the action of G on V has property (F), and one may take for V' the space V with the linear varieties $\ker(g-1)$ removed, for g in $G - \{1\}$.

5.6. EXAMPLES.

(1) Choose $G = \mathbf{O}_n$, so that $A = \mathrm{Quad}_n$, and assume that $\mathrm{char}(k) \neq 2$. Take $K = k_0(t_1, t_2, \ldots, t_n)$ where the t_i are indeterminates. The torsor corresponding to the quadratic form $\langle t_1, \ldots, t_n \rangle$ is versal.

The density property 5.1.2 holds in this case because the class of the quadratic form stays the same when one multiplies the $t_1, \ldots, t_n$ by arbitrary squares, and such changes give rise to a dense set in the space of the t_i's. A similar remark applies to practically all of the versal objects that appear in these notes (e.g., for G_2 or F_4); in each case, there are trivial changes of the parameters which give a dense set.

(2) $G = S_n$, so $A = \mathrm{Et}_n$. Take $K = k_0(c_1, \ldots, c_n)$ where the c_i are indeterminates. Define the K-étale algebra:

$$E = K[x]/(x^n + c_1 x^{n-1} + \cdots + c_n).$$

The corresponding torsor is versal; it is the special case of 5.5 where V is affine n-space and S_n acts by permuting the coordinates, see 24.6.

5.7. ESSENTIAL DIMENSION. The minimal transcendence degree of K/k_0 for K the field of definition of a versal G-torsor in 5.1 is called the *essential dimension* of G, cf. [Re 00]. This is an interesting invariant of G, whose explicit determination is not easy in general (even if G is finite). Here are some examples (assuming the characteristic is $\neq 2$):

$$\mathrm{ed}(\mathbf{O}_n) = n \quad \text{for } n \geq 1$$
$$\mathrm{ed}(\mathbf{SO}_n) = n - 1 \quad \text{for } n \geq 3$$
$$\mathrm{ed}(G_2) = 3.$$

The cyclic group of order n has essential dimension 1 if k_0 contains a primitive n-th root of unity. However, the cyclic group of order 4 has essential dimension 2 if k_0 does not contain a primitive 4th root of unity.[e] Many more examples may be found in [Re 00].

REMARK 5.8. One could also broaden the notion of versal torsor by removing the density condition in 5.1.2; that is, by replacing 5.1.2 with the following:

(2′) For every extension k of k_0 with k infinite and every G-torsor T over k, there exists $x \in X(k)$ whose fiber Q_x is isomorphic to T.

The results here — specifically, 12.3 and 27.13 — still hold with this broader definition.

[e]See [Se 92b, p. 6, exercise]. Note a misprint in this exercise: $+6Z^2$ should twice be replaced with $-6Z^2$.

CHAPTER II

Cohomological preliminaries: the local case

The results of this chapter and of the next one are essentially standard.[a] The basic idea of Section 9 originates with [Fa 51]. Much of §§6, 7, and 9 can be found in [Ar 75] when $C = \mathbf{Z}/2\mathbf{Z}$ and k has characteristic $\neq 2$ and in [El 82] when $C = \mathbf{Z}/n\mathbf{Z}$ and k contains a primitive n-th root of unity. A résumé of the general case was given in [Se 92a] and reproduced in the fifth edition of [Se 65].

6. An exact sequence

Let G be a profinite group and N a (closed) normal subgroup of G. Set $\Gamma = G/N$, and let C be a discrete Γ-module, which we also view as a discrete G-module with trivial action by N.

THEOREM 6.1. *Suppose that*

(1) $H^i(N, C) = 0$ *if* $i > 1$.

(2) *The exact sequence* $1 \to N \to G \to \Gamma \to 1$ *splits.*

Then for all $i \geq 0$*, we have an exact sequence*

$$0 \to H^i(\Gamma, C) \xrightarrow{\pi} H^i(G, C) \xrightarrow{r} H^{i-1}(\Gamma, \mathrm{Hom}(N, C)) \to 0,$$

where π *is the natural map defined by* $G \to \Gamma$ *and* r *is as described below.*

[Here, $\mathrm{Hom}(N, C)$ denotes the group of *continuous* homomorphisms $N \to C$ (i.e., those with open kernel and hence finite image).]

PROOF. The spectral sequence of group extensions, as constructed in [HS 53], gives

$$H^p(\Gamma, H^q(N, C)) \Rightarrow H(G, C).$$

We have $H^0(N, C) = C$ and $H^1(N, C) = \mathrm{Hom}(N, C)$. By hypothesis (1), we also have $H^i(N, C) = 0$ for $i > 1$, hence the spectral sequence reduces to a long exact sequence

$$\cdots \to H^i(\Gamma, C) \xrightarrow{\pi} H^i(G, C) \xrightarrow{r} H^{i-1}(\Gamma, \mathrm{Hom}(N, C)) \to$$
$$\xrightarrow{d} H^{i+1}(\Gamma, C) \to H^{i+1}(G, C) \to \cdots .$$

Here d is the "d_2-differential" of the spectral sequence. Hypothesis (2) implies that π is injective, hence $d = 0$, so that the exact sequence above splits into short exact sequences

$$0 \to H^i(\Gamma, C) \xrightarrow{\pi} H^i(G, C) \xrightarrow{r} H^{i-1}(\Gamma, \mathrm{Hom}(N, C)) \to 0.$$

[a] It would be a good exercise for the reader to deduce the content of these two chapters from the general machinery of étale cohomology. The main difficulty is the comparison between the residues given by the étale theory and the residues defined in the present text; they may differ by a sign. Which one?

This proves the theorem.

As for the explicit definition of r, it is as follows: By [HS 53, p. 121], an element $\alpha \in H^i(G, C)$ can be represented by a cocycle $f(g_1, \ldots, g_i)$ which is normalized (i.e., is equal to 0 when one of the g_j is equal to 1) and which only depends on the image of g_j in Γ for $j \geq 2$. For $\gamma_1, \ldots, \gamma_{i-1}$ in Γ, let $r(f)(\gamma_1, \ldots, \gamma_{i-1})$ be the element of $\mathrm{Hom}(N, C)$ defined by

$$r(f)(\gamma_1, \ldots, \gamma_{i-1})(n) = f(n, g_1, \ldots, g_{i-1})$$

where the g_j are representatives of the γ_j in G and n runs through the elements of N. The cochain $r(f)$ is an $(i-1)$-cocycle of Γ with values in $\mathrm{Hom}(N, C)$. The class of $r(f)$ in $H^{i-1}(\Gamma, \mathrm{Hom}(N, C))$ is independent of the choice of f; it is $r(\alpha)$. □

REMARK 6.2. When $i = 0$, $H^{i-1}(\Gamma, \mathrm{Hom}(N, C))$ is to be interpreted as 0.

When $i = 1$, the map $r\colon H^1(G, C) \to H^0(\Gamma, \mathrm{Hom}(N, C)) = \mathrm{Hom}(N, C)^\Gamma$ is the restriction map $H^1(G, C) \to H^1(N, C) = \mathrm{Hom}(N, C)$.

REMARK 6.3. Beware that different constructions of "the" spectral sequence of group extensions (see [Be 81]) could lead to an r map different from ours by the sign $(-1)^{i-1}$, and then Proposition 6.6 would also involve a sign. Such sign problems are one of the reasons why Galois cohomology is being used here instead of the (more powerful) étale cohomology.

REMARK 6.4. The r-map is obviously functorial: A commutative diagram

$$\begin{array}{ccccccccc} 1 & \longrightarrow & N' & \longrightarrow & G' & \longrightarrow & \Gamma' & \longrightarrow & 1 \\ & & \downarrow & & \downarrow & & \downarrow & & \\ 1 & \longrightarrow & N & \longrightarrow & G & \longrightarrow & \Gamma & \longrightarrow & 1 \end{array}$$

gives rise to a diagram

$$\begin{array}{ccc} H^i(G', C) & \longrightarrow & H^{i-1}(\Gamma', \mathrm{Hom}(N', C)) \\ \uparrow & & \uparrow \\ H^i(G, C) & \longrightarrow & H^{i-1}(\Gamma, \mathrm{Hom}(N, C)) \end{array}$$

which is commutative. (Assuming, of course, that $H^i(N, C)$ and $H^i(N', C)$ are 0 for all $i > 1$.)

6.5. CUP PRODUCT FORMULAS. Let C_1, C_2, C_3 be Γ-modules, with a pairing $C_1 \times C_2 \to C_3$. Assume the hypotheses of the theorem for C_1 and C_3. Let $\alpha_1 \in H^{i_1}(G, C_1)$ and $\alpha_2 \in H^{i_2}(\Gamma, C_2) \subset H^{i_2}(G, C_2)$. Let $\alpha_1 \cdot \alpha_2 \in H^{i_1+i_2}(G, C_3)$ be the cup product of α_1 and α_2 relative to the given pairing.

PROPOSITION 6.6. *We have* $r(\alpha_1 \cdot \alpha_2) = r(\alpha_1) \cdot \alpha_2$ *in* $H^{i_1+i_2-1}(\Gamma, \mathrm{Hom}(N, C_3))$.

(We have $r(\alpha_1) \in H^{i_1-1}(\Gamma, \mathrm{Hom}(N, C_1))$ and $\alpha_2 \in H^{i_2}(\Gamma, C_2)$; the cup product $r(\alpha_1) \cdot \alpha_2$ is relative to the natural pairing $\mathrm{Hom}(N, C_1) \times C_2 \to \mathrm{Hom}(N, C_3)$.)

PROOF. Choose a normalized cocycle f_1 representing α_1 of the type described in the proof of Th. 6.1. Let f_2 be a normalized Γ-cocycle representing α_2, which we view also as a G-cocycle. Then $\alpha_1 \cdot \alpha_2$ can be represented by the cup product $f_1 \cdot f_2$; recall that

$$(f_1 \cdot f_2)(g_1, \ldots, g_{i_1+i_2}) = f_1(g_1, \ldots, g_{i_1}) \cdot {}^{g_1 \cdots g_{i_1}} f_2(g_{i_1+1}, \ldots, g_{i_1+i_2}).$$

A direct computation shows that $r(f_1 \cdot f_2) = r(f_1) \cdot f_2$. The proposition follows. □

7. Galois cohomology of local fields

7.1. INERTIA GROUPS. Let K be a field with a discrete valuation v with residue field $k(v) = k$. Let K_s be a separable closure of K, and put $\Gamma_K = \mathrm{Gal}(K_s/K)$ and similarly for Γ_k. Assume until 7.13 that K is *complete*. The valuation v extends uniquely to K_s; its residue field is an algebraic closure of k. This gives a surjection $\Gamma_K \to \Gamma_k$ with kernel the *inertia group* I.

We have a natural exact sequence

$$1 \to I_{\mathrm{wild}} \to I \to I_t \to 1,$$

where

$$I_t = \text{tame inertia group (see below)}$$

and

$$I_{\mathrm{wild}} = \begin{cases} 1 & \text{if } \mathrm{char}(k) = 0 \\ \text{the unique } p\text{-Sylow subgroup of } I & \text{if } \mathrm{char}(k) = p. \end{cases}$$

Let K_t be the subfield of K_s fixed by I_{wild}; it has residue field k_s. Let K_{unr} be the subfield of K_s consisting of the elements fixed by I. Then K_t is generated as a K_{unr}-algebra by all n-th roots of a uniformizing element π of K for n not divisible by the residue characteristic. For such n, let $\boldsymbol{\mu}_n$ be the group of n-th roots of unity (in k_s^* or K_{unr}^*, it amounts to the same). We have an isomorphism $I_t \to \varprojlim \boldsymbol{\mu}_n$ given by sending $s \in I_t$ to $\zeta_s \in \boldsymbol{\mu}_n$ defined by $\zeta_s = s(\pi^{1/n})/\pi^{1/n}$. It does not depend on the choice of π nor on that of $\pi^{1/n}$.

REMARK 7.2. The above description of I and I_t gives canonical isomorphisms:

$$\mathrm{Hom}(I, \boldsymbol{\mu}_n) = \mathrm{Hom}(I_t, \boldsymbol{\mu}_n) = \mathbf{Z}/n\mathbf{Z}. \tag{7.3}$$

If one identifies $\mathrm{Hom}(I, \boldsymbol{\mu}_n)$ with $K_{\mathrm{unr}}^*/K_{\mathrm{unr}}^{*n}$ by Kummer theory, the isomorphism $\mathrm{Hom}(I, \boldsymbol{\mu}_n) = \mathbf{Z}/n\mathbf{Z}$ becomes the map

$$K_{\mathrm{unr}}^*/K_{\mathrm{unr}}^{*n} \to \mathbf{Z}/n\mathbf{Z} \quad \text{given by } x \mapsto v(x) \pmod{n}.$$

7.4. COHOMOLOGY. For the rest of this section, C denotes a finite Γ_k-module of order not divisible by the residue characteristic. We also view C as a Γ_K-module with trivial action by I. (Most of what follows also works for infinite torsion modules when their elements have order prime to the residue characteristic.)

LEMMA 7.5. *$H^i(I, C) = 0$ for $i \geq 2$.*

PROOF. The ℓ-Sylow subgroups of I (for $\ell \neq \mathrm{char}(k)$) are isomorphic to $\mathbf{Z}_\ell$. Hence the ℓ-cohomological dimension of I is 1; the lemma follows. □

LEMMA 7.6. *The sequence $1 \to I \to \Gamma_K \to \Gamma_k \to 1$ is split.*

PROOF. (cf. [Ax 65] and [Se 65, II.4.3, exerc.]) For each $m > 1$ and not divisible by the residue characteristic, choose an m-th root π_m of π so that the choices are compatible with taking powers (i.e., $(\pi_{mm'})^{m'} = \pi_m$). The extension K' of K generated by the π_m's is linearly disjoint from K_{unr} and we have $K_{\mathrm{unr}} K' = K_t$. This shows that the exact sequence $1 \to I_t \to \Gamma_K/I_{\mathrm{wild}} \to \Gamma_k \to 1$ splits. This proves the lemma if $\mathrm{char}(k) = 0$. If $\mathrm{char}(k) = p > 0$, the p-cohomological dimension of Γ_k is ≤ 1 [Se 65, II.2.2, Prop. 3], so that every homomorphism $\Gamma_k \to \Gamma_K/I_{\mathrm{wild}}$ can be lifted to Γ_K [Se 65, I.3.4, Prop. 16]. □

PROPOSITION 7.7. *The sequence*

$$0 \to H^i(\Gamma_k, C) \to H^i(\Gamma_K, C) \xrightarrow{r} H^{i-1}(\Gamma_k, \mathrm{Hom}(I, C)) \to 0$$

is exact.

(For the definition of r, see 6.1.)

This follows from Theorem 6.1 applied to $G = \Gamma_K$, $N = I$, and $\Gamma = \Gamma_k$.

7.8. TATE TWISTS. Let us choose a number n not divisible by the residue characteristic such that $nC = 0$. If d is an integer, we define the d-th *Tate twist* of C to be

$$C(d) = \begin{cases} \boldsymbol{\mu}_n^{\otimes d} \otimes C & \text{if } d \geq 0 \\ \mathrm{Hom}(\boldsymbol{\mu}_n^{\otimes -d}, C) & \text{if } d < 0. \end{cases}$$

This does not depend on the particular choice of n, and this definition works over any ground field. (In particular, the ground field need not be local.)

For example, $C(-1) = \mathrm{Hom}(\boldsymbol{\mu}_n, C)$.

The Tate twist $C(d)$ can also be defined when C is not assumed to be finite, but merely torsion without any element of order equal to the characteristic. One writes C as $\varinjlim C'$, where the C' are the finite submodules of C, and one defines $C(d)$ as $\varinjlim C'(d)$. In particular, $C(-1) = \mathrm{Hom}(M, C)$, where $M = \varprojlim \boldsymbol{\mu}_n$ and Hom denotes continuous homomorphisms.

7.9. RESIDUES. For K as in 7.1, $\mathrm{Hom}(I, C)$ is the same as $\mathrm{Hom}(I_t, C)$, hence the same as $\mathrm{Hom}(\boldsymbol{\mu}_n, C)$. The exact sequence 7.7 can be rewritten as

$$(7.10) \qquad 0 \to H^i(k, C) \to H^i(K, C) \xrightarrow{r} H^{i-1}(k, C(-1)) \to 0.$$

For $\alpha \in H^i(K, C)$, $r(\alpha)$ is called the *residue* of α.

7.11. For $x \in K^*$, write $(x)_n$ for the corresponding element in $K^*/K^{*n} = H^1(K, \boldsymbol{\mu}_n)$.

PROPOSITION. *Let π be a uniformizing element of K. Every $\alpha \in H^i(K, C)$ can be written uniquely as:*

$$\alpha = \alpha_0 + (\pi)_n \cdot \alpha_1 \textit{ with } \alpha_0 \in H^i(k, C) \textit{ and } \alpha_1 \in H^{i-1}(k, C(-1)).$$

Moreover, we have $r(\alpha) = \alpha_1$.

PROOF. The cup product $(\pi)_n \cdot \alpha_1$ is the one induced by the natural pairing $\boldsymbol{\mu}_n \times C(-1) \to C$. The residue map

$$r \colon H^1(K, \boldsymbol{\mu}_n) \to H^0(k, \mathbf{Z}/n\mathbf{Z})$$

sends $(x)_n$ to $v(x) \in \mathbf{Z}/n\mathbf{Z}$ by Remark 6.2. By 6.5, $r((\pi_n) \cdot \alpha_1) = \alpha_1$. This shows the uniqueness of any decomposition of α as $\alpha = \alpha_0 + (\pi)_n \cdot \alpha_1$. The existence follows from the fact that, if α_1 is defined as $r(\alpha)$, then the residue of $\alpha - (\pi)_n \cdot \alpha_1$ is zero. □

EXERCISE 7.12 (residue of a cup product). Let A, B, C be finite Γ_k-modules with a pairing $A \times B \to C$. Let $n > 0$ be prime to char(k) and such that $nA = 0$, $nB = 0$, and $nC = 0$. Let $\alpha \in H^a(K, A)$, $\beta \in H^b(K, B)$, and let $\alpha \cdot \beta$ denote their cup product in $H^{a+b}(K, C)$. Prove the formula:

$$r(\alpha \cdot \beta) = r(\alpha) \cdot \beta + (-1)^a \, \alpha \cdot r(\beta) + r(\alpha) \cdot r(\beta) \cdot (-1)_n,$$

where $r(\alpha)$, $r(\beta)$, $r(\alpha\cdot\beta)$ are viewed as elements of the cohomology of K, via the natural embedding $H(k) \to H(K)$. The cup products are taken relatively to the natural pairings; for instance, the last one is relative to

$$A(-1) \times B(-1) \times \boldsymbol{\mu}_n \to C(-2) \times \boldsymbol{\mu}_n \to C(-1).$$

All the terms of the formula thus belong to $H^{a+b-1}(K, C(-1))$.

[Hint: Decompose α and β as $\alpha_0 + (\pi)_n \cdot \alpha_1$ and $\beta_0 + (\pi)_n \cdot \beta_1$, as in Prop. 7.11, and use the identity $(\pi)_n \cdot (-\pi)_n = 0$.]

7.13. THE NON-COMPLETE CASE. We now drop the assumption that K is complete with respect to v. Choose an extension w of v to K_s (all such are Γ_K-conjugate), and let Dec_w be the corresponding decomposition group, which consists of those $g \in \Gamma_K$ such that $gw = w$.

Let $K_{s,w}$ be the completion of K_s relative to w. The subfield $K_s \cdot K_v$ of $K_{s,w}$ is the largest algebraic subextension of $K_{s,w}/K_v$; it is separably closed by Krasner's Lemma, hence may be identified with $(K_v)_s$. We have

$$\mathrm{Dec}_w = \mathrm{Gal}(K_s \cdot K_v / K_v) = \mathrm{Gal}((K_v)_s / K_v) = \Gamma_{K_v},$$

so that we may apply the results above to Dec_w.

More precisely, let C be a finite Γ_K-module of order not divisible by $\mathrm{char}(k)$. Assume that C is *unramified at* v, i.e., that the inertia group of Dec_w acts trivially on C. Let $\alpha \in H^i(K, C)$, and let α_v denote its image in $H^i(K_v, C) = H^i(\mathrm{Dec}_w, C)$. We define the *residue* $r_v(\alpha)$ *of* α *at* v to be the residue of α_v in $H^{i-1}(k, C(-1))$ as defined above. If $r_v(\alpha) \neq 0$, we say that α is *ramified* at v (or has a *pole* at v). If $r_v(\alpha) = 0$, α is said to be *unramified* at v. In that case, α_v can be identified with an element of $H^i(k, C)$ which we denote by $\alpha(v)$ and call the *value of* α *at* v.

We thus have two canonical maps relating the cohomology of K with that of the residue field k of v:

$$r_v\colon H^i(K, C) \to H^{i-1}(k, C(-1)) \qquad \text{(residue at } v\text{)}$$
$$\ker(r_v) \to H^i(k, C) \qquad \text{(value at } v\text{)}.$$

When K is complete, we have seen that the first map is surjective, and that the second one is an isomorphism. This is still true when K is henselian, but not in general.

8. Functoriality of the restriction and corestriction

8.1. RESTRICTION. Recall the setting of 7.1 and 7.13: (K, v) is a field with a discrete valuation v and residue field k; we do not assume it is complete. We write C for a finite Γ_K-module which is unramified at v and of order prime to the characteristic of k.

Let (K', v') be an extension of (K, v) with ramification index e and residue field k'. We have residue maps

$$r_v\colon H^i(K, C) \to H^{i-1}(k, C(-1)) \quad \text{and} \quad r_{v'}\colon H^i(K', C) \to H^{i-1}(k', C(-1)).$$

PROPOSITION 8.2. (1) *The diagram*

$$\begin{array}{ccc} H^i(K, C) & \xrightarrow{r_v} & H^{i-1}(k, C(-1)) \\ \downarrow & & \downarrow e \\ H^i(K', C) & \xrightarrow{r_{v'}} & H^{i-1}(k', C(-1)) \end{array}$$

is commutative, where the vertical arrow on the right is the product of e *and the natural map* $H^{i-1}(k, C(-1)) \to H^{i-1}(k', C(-1))$.

(2) *The diagram*

$$\begin{array}{ccc} \ker(r_v) & \longrightarrow & H^i(k,C) \\ \downarrow & & \downarrow \\ \ker(r_{v'}) & \longrightarrow & H^i(k',C) \end{array}$$

is commutative.

(For the definition of the map $\ker(r_v) \to H^i(k, C)$, see 7.13.)

PROOF. We may assume that K and K' are complete. Let $I_t(K)$ and $I_t(K')$ be the corresponding tame inertia groups. By 7.1, we have a natural isomorphism of $I_t(K)$ with $\varprojlim \boldsymbol{\mu}_n$. This identification used a uniformizing element of K, and changing K to K' forces a new choice of uniformizing element. The result is a commutative diagram

$$\begin{array}{ccc} I_t(K) & = & \varprojlim \boldsymbol{\mu}_n \\ \uparrow & & \uparrow{\scriptstyle e} \\ I_t(K') & = & \varprojlim \boldsymbol{\mu}_n, \end{array} \tag{8.3}$$

where the vertical map on the left is the natural map and the vertical map on the right is multiplication by the ramification index e. By 7.9, we have $C(-1) = \mathrm{Hom}(I_t, C)$, where I_t may be taken to be either inertia group. Via these identifications, the natural map $\mathrm{Hom}(I_t(K), C) \to \mathrm{Hom}(I_t(K'), C)$ becomes the map $C(-1) \to C(-1)$ given by multiplication by e. It follows from 6.4 that one has a commutative diagram

$$\begin{array}{ccccccccc} 0 & \longrightarrow & H^i(k,C) & \longrightarrow & H^i(K,C) & \longrightarrow & H^{i-1}(k,C(-1)) & \longrightarrow & 0 \\ & & \downarrow & & \downarrow & & \downarrow{\scriptstyle e} & & \\ 0 & \longrightarrow & H^i(k',C) & \longrightarrow & H^i(K',C) & \longrightarrow & H^{i-1}(k',C(-1)) & \longrightarrow & 0, \end{array} \tag{8.4}$$

where the downward arrow on the right is given by the product of the ramification index e with the natural homomorphism $H^{i-1}(k, C(-1)) \to H^{i-1}(k', C(-1))$. Prop. 8.2 follows. □

8.5. CORESTRICTION. Suppose now that K' is finite over K, and that K (and hence also K') is complete. Then, one has a *corestriction* map

$$\mathrm{Cor}_K^{K'} : H^i(K', C) \to H^i(K, C)$$

defined to be the product of the usual Galois corestriction (corresponding to the inclusion $\Gamma_{K'} \to \Gamma_K$) with the degree of inseparability $[K' : K]_i$ of K' over K. (This definition applies to any finite field extension, for instance to the residue field extension k'/k.)

PROPOSITION 8.6. *The diagram*

$$\begin{array}{ccccccccc} 0 & \longrightarrow & H^i(k,C) & \longrightarrow & H^i(K,C) & \longrightarrow & H^{i-1}(k,C(-1)) & \longrightarrow & 0 \\ & & \uparrow{\scriptstyle e.\operatorname{Cor}_k^{k'}} & & \uparrow{\scriptstyle \operatorname{Cor}_K^{K'}} & & \uparrow{\scriptstyle \operatorname{Cor}_k^{k'}} & & \\ 0 & \longrightarrow & H^i(k',C) & \longrightarrow & H^i(K',C) & \longrightarrow & H^{i-1}(k',C(-1)) & \longrightarrow & 0, \end{array}$$

commutes, where the upward arrow on the left is given by the product of the ramification index e with $\operatorname{Cor}_k^{k'}$.

PROOF. It is sufficient to prove the proposition in the following two special cases:

CASE I: k'/k is purely inseparable.
CASE II: $e = 1$ and k'/k is separable (the "unramified case").

Suppose that we have proved the result in both cases, and consider the general case. There is a unique field k'' with $k \subset k'' \subset k'$ such that k''/k is separable and k'/k'' is purely inseparable. It is well-known that there is a field K'' with $K \subset K'' \subset K'$ such that $e(K''/K) = 1$ and the residue field of K'' is k'', see e.g. [Bou 83, App.]. The extension K''/K is of the sort occurring in Case II and the extension K'/K'' is as in Case I. Hence, if the proposition holds in each of these two special cases, then it holds in general.

Consider first the left-hand square

$$\begin{array}{ccc} H^i(k,C) & \longrightarrow & H^i(K,C) \\ \uparrow{\scriptstyle e.\operatorname{Cor}_k^{k'}} & & \uparrow{\scriptstyle \operatorname{Cor}_K^{K'}} \\ H^i(k',C) & \longrightarrow & H^i(K',C) \end{array} \tag{8.7}$$

of the diagram in 8.6.

8.8. DIAGRAM (8.7) COMMUTES IN CASE I. In Case I, k' is purely inseparable over k and we have $\Gamma_{k'} = \Gamma_k$. Thus the restriction map $H^i(k,C) \to H^i(k',C)$ is surjective (even an isomorphism), and it is enough to prove the commutativity of

$$\begin{array}{ccc} H^i(k,C) & \longrightarrow & H^i(K,C) \\ \uparrow{\scriptstyle e.\operatorname{Cor}_k^{k'}} & & \uparrow{\scriptstyle \operatorname{Cor}_K^{K'}} \\ H^i(k',C) & \longrightarrow & H^i(K',C) \\ \uparrow & & \uparrow \\ H^i(k,C) & \longrightarrow & H^i(K,C). \end{array}$$

(Here the lower vertical maps are the restriction maps.) The composition of the vertical arrows on the right is multiplication by $n = [K' : K]$. The composition of the vertical arrows on the left is e times multiplication by $f = [k' : k]$. Since $ef = n$, the square obtained by omitting the middle row commutes. Since the bottom square commutes and the vertical map in the lower left is a surjection, the top square — which is the square in (8.7) — commutes.

8.9. DIAGRAM (8.7) COMMUTES IN CASE II. In Case II, the natural map $\Gamma_K \to \Gamma_k$ gives a bijection of $\Gamma_K/\Gamma_{K'}$ onto $\Gamma_k/\Gamma_{k'}$. In this situation, it is clear that Diagram (8.7) commutes.

REMARK 8.10. The only property of C used in 8.8 and 8.9 was that it is a Γ_k-module. Hence Diagram (8.7) commutes for C any unramified module; C need not be finite of order prime to the residue characteristic.

To conclude the proof of Prop. 8.6 we must show that the right-hand square

$$\begin{array}{ccc} H^i(K,C) & \xrightarrow{r_K} & H^{i-1}(k,C(-1)) \\ {\scriptstyle \mathrm{Cor}_K^{K'}}\uparrow & & \uparrow{\scriptstyle \mathrm{Cor}_k^{k'}} \\ H^i(K',C) & \xrightarrow{r_{K'}} & H^{i-1}(k',C(-1)) \end{array} \tag{8.11}$$

commutes, where the horizontal arrows are the residue maps.

Start with an element x' of $H^i(K',C)$. We want to compute the residue of $y = \mathrm{Cor}_K^{K'}(x')$. By Proposition 7.11 we may write x' as $x'_0 + (\pi')\cdot x'_1$ where π' is a uniformizing element of K', x'_0 is in $H^i(k',C)$, and x'_1 is in $H^{i-1}(k',C(-1))$ such that $x'_1 = r_{K'}(x')$. By the commutativity of (8.7), the corestriction $\mathrm{Cor}(x'_0)$ belongs to the subgroup $H^i(k,C)$ of $H^i(K,C)$, and its residue is 0. We may then assume that $x' = (\pi')\cdot x'_1$.

Let us consider separately the cases I and II.

8.12. DIAGRAM (8.11) COMMUTES IN CASE I. In Case I, the restriction map $H^{i-1}(k,C(-1)) \to H^{i-1}(k',C(-1))$ is surjective. Hence we may write x'_1 as a restriction $\mathrm{Res}_k^{k'}(x_1)$, where x_1 is an element of $H^{i-1}(k,C(-1))$. There is a "projection formula" which gives $\mathrm{Cor}(x\cdot y)$ where x (or y) is $\mathrm{Res}(X)$ (or $\mathrm{Res}(Y)$), namely

$$\mathrm{Cor}(x\cdot y) = X\cdot \mathrm{Cor}(y) \quad \text{or} \quad \mathrm{Cor}(x)\cdot Y, \tag{8.13}$$

see e.g. [Lang 96, p. 87, Th. 3.1(2)] or [NSW 00, p. 47, Prop. 1.5.3.(iv)]. Hence we have

$$y = \mathrm{Cor}((\pi')\cdot x'_1) = \mathrm{Cor}((\pi'))\cdot x_1.$$

It is a standard fact that $\mathrm{Cor}((z)_n) = (Nz)_n$, where N is the norm. Hence we get $y = (N\pi')\cdot x_1$, and its residue is equal to $v(N\pi')\cdot x_1 = f\,x_1$, where $f = [k' : k]$. Since $f\,x_1$ is the corestriction of x_1 with respect to the purely inseparable extension k'/k, we get the formula we wanted.

8.14. DIAGRAM (8.11) COMMUTES IN CASE II. In Case II, we may take for π' a uniformizing element π of K. By the projection formula (8.13), we have:

$$\mathrm{Cor}_K^{K'}(x') = (\pi)\cdot \mathrm{Cor}_k^{k'}(x'_1).$$

This shows that the residue of $\mathrm{Cor}(x')$ is equal to $\mathrm{Cor}(x'_1)$, as desired.

This completes the proof of Proposition 8.6. □

CHAPTER III

Cohomological preliminaries: the function field case

9. Galois cohomology of $k(t)$

Let $\mathbb{P}_1 = \mathbb{P}_{1/k}$ be the projective line over k. The function field of $\mathbb{P}_1$ is $K = k(t)$ where t is an indeterminate. A closed point of $\mathbb{P}_1$ can be identified with a discrete valuation on K which is trivial on k. Let V be the set of such valuations. Suppose that C is a finite Γ_k-module whose order n is not divisible by the characteristic of k. As above, for every $v \in V$ we call K_v the completion of K at v; it is a local field; its residue field $k(v)$ is a finite extension of k. We may apply the definitions of 7.13; in particular we have a residue map

$$r_v \colon H^i(K, C) \to H^{i-1}(k(v), C(-1)).$$

LEMMA 9.1. *Let α be an element of $H^i(K, C)$. The set of v in V which are poles of α (in the sense of 7.13) is finite.*

PROOF. Let L/K be a finite Galois extension of K which is large enough so that α comes from an element of $H^i(\mathrm{Gal}(L/K), C_L)$, where $C_L = H^0(\Gamma_L, C)$. Then $r_v(\alpha) = 0$ for all v where L is unramified. (It would be sufficient that the ramification index of L/K is prime to the order of C.) This follows from the fact that the kernel of the residue map is $H^i(\mathrm{Gal}(K_{v,\mathrm{unr}}/K_v), C)$. □

This allows us to define a map

$$\oplus r_v \colon H^i(K, C) \to \oplus_{v \in V} H^{i-1}(k(v), C(-1))$$

by $\alpha \mapsto (r_v(\alpha))$. We have the following basic result, which will be proved in 9.5 through 9.20.

THEOREM 9.2. *The sequence*

$$0 \to H^i(k, C) \to H^i(K, C) \xrightarrow{\oplus r_v} \bigoplus_{v \in V} H^{i-1}(k(v), C(-1)) \xrightarrow{c} H^{i-1}(k, C(-1)) \to 0$$

is exact, where c is the direct sum of the corestriction maps

$$\mathrm{Cor} \colon H^{i-1}(k(v), C(-1)) \to H^{i-1}(k, C(-1)).$$

An element of $H^i(K, C)$ is said to be *constant* if it belongs to $H^i(k, C)$.

The fact that the composition $c \circ (\oplus r_v)$ is zero is known as the "residue formula"; it says that the sum of the corestrictions of the residues of an element of $H^i(K, C)$ is zero.

Note that the point at infinity of $\mathbb{P}_1$ defines a place, which we denote by ∞, with residue field $k(\infty) = k$. The corresponding corestriction is the identity. Hence we may remove $H^{i-1}(k, C(-1))$ from the last two groups in the exact sequence in 9.2. This gives:

THEOREM 9.3. *The sequence*

$$0 \to H^i(k,C) \to H^i(K,C) \to \bigoplus_{v\in V-\{\infty\}} H^{i-1}(k(v),C(-1)) \to 0$$

is exact.

9.4. UNRAMIFIED ELEMENTS. An element $\alpha \in H^i(K,C)$ is called *unramified* (resp. *unramified outside* 0, ∞) if it has zero residue for all $v \in V$ (resp. $v \in V - \{0,\infty\}$). Because of the residue formula, α is unramified if it has zero residue for all $v \in V-\{\infty\}$, i.e., on the affine line. Write $H^i_{\text{unr}}(K,C)$ (resp. $H^i_{\text{unr. outside } 0,\infty}(K,C)$) for the group of such classes. Theorem 9.3 shows that any unramified α belongs to $H^i(k,C)$, i.e., $H^i_{\text{unr}}(K,C) = H^i(k,C)$.

If $\alpha \in H^i_{\text{unr. outside } 0,\infty}(K,C)$ has residue $\alpha_1 \in H^{i-1}(k,C(-1))$ at 0, then $\alpha - (t)_n{\cdot}\alpha_1$ is unramified by 7.11. Hence such an α can be written uniquely as $\alpha_0 + (t)_n{\cdot}\alpha_1$, with $\alpha_0 \in H^i(k,C)$ and $\alpha_1 \in H^{i-1}(k,C(-1))$. By comparing with Proposition 7.11, this shows that the natural map

$$H^i_{\text{unr. outside } 0,\infty}(K,C) \to H^i(K_{v_0},C)$$

is an isomorphism, where $K_{v_0} = k((t))$ is the completion of K at the place 0.

9.5. PROOF OF THEOREM 9.2. We assume first that k is *perfect* (for the general case, see 9.18), so that $k_s(t) = \overline{k}(t)$. Let $N = \text{Gal}(K_s/\overline{k}(t))$. It is well-known (cf. [Se 65, II.3]) that the cohomological dimension of $\overline{k}(t)$ is 1, so $H^i(N,C) = 0$ for all $i \geq 2$.

For $v \in V$ choose an extension w to K_s and let I_w and Dec_w denote the inertia and decomposition groups of w, which are subgroups of Γ_K. The diagram

$$\begin{array}{ccccccccc} 1 & \longrightarrow & I_w & \longrightarrow & \text{Dec}_w & \longrightarrow & \Gamma_{k(v)} & \longrightarrow & 1 \\ & & {\scriptstyle i_w}\downarrow & & \downarrow & & \downarrow & & \\ 1 & \longrightarrow & N & \longrightarrow & \Gamma_K & \longrightarrow & \Gamma_k & \longrightarrow & 1 \end{array} \tag{9.6}$$

commutes, where the vertical maps are the obvious ones. If one chooses v such that $k(v) = k$ (e.g., $v = \infty$), by 7.6 and 7.13 the upper sequence splits, hence the lower one does as well.

We have checked all of the hypotheses of Theorem 6.1 for the exact sequence $1 \to N \to \Gamma_K \to \Gamma_k \to 1$. We thus have an exact sequence

$$0 \to H^i(k,C) \to H^i(K,C) \xrightarrow{r} H^{i-1}(k,\text{Hom}(N,C)) \to 0.$$

It remains to show that

$$H^{i-1}(k,\text{Hom}(N,C)) \;\cong\; \ker\left(\oplus H^{i-1}(k(v),C(-1)) \xrightarrow{c} H^{i-1}(k,C(-1))\right), \tag{9.7}$$

and

(9.8) *the v-component of r is the residue map r_v.*

We do both by looking more closely at the structure of $\text{Hom}(N,\boldsymbol{\mu}_n)$ and $\text{Hom}(N,C)$.

9.9. STRUCTURE OF $\text{Hom}(N,\boldsymbol{\mu}_n)$. Let

$$\overline{V} = \mathbb{P}_1(\overline{k}) = \{\text{discrete valuations of } \overline{k}(t)\},$$

and let D be the group of divisors of $\mathbb{P}_1$ over $\overline{k}$, i.e., the free $\mathbf{Z}$-module with basis $\overline{V}$. Let D_0 be the subgroup of D made up of divisors of degree 0. We have an exact sequence

$$0 \to D_0/nD_0 \to D/nD \xrightarrow{\deg} \mathbf{Z}/n\mathbf{Z} \to 0 \tag{9.10}$$

If f is an element of $\overline{k}(t)^*$, the divisor $\operatorname{div}(f)$ of f belongs to D_0; the map $f \mapsto \operatorname{div}(f)$ induces an isomorphism

$$\overline{k}(t)^*/\overline{k}(t)^{*n} \xrightarrow{\sim} D_0/nD_0. \tag{9.11}$$

By Kummer theory, we also have a natural isomorphism

$$\operatorname{Hom}(N, \boldsymbol{\mu}_n) = \overline{k}(t)^*/\overline{k}(t)^{*n}. \tag{9.12}$$

By combining (9.11) and (9.12), we obtain:

$$\operatorname{Hom}(N, \boldsymbol{\mu}_n) = D_0/nD_0. \tag{9.13}$$

Now let v be an element of V, and let w be an extension of v to K_s. The restriction of w to $\overline{k}(t)$ defines an element $\overline{v}$ of $\overline{V}$. The map $i_w : I_w \to N$ of (9.6) gives

$$i_w^* : \operatorname{Hom}(N, \boldsymbol{\mu}_n) \to \operatorname{Hom}(I_w, \boldsymbol{\mu}_n) = \mathbf{Z}/n\mathbf{Z}. \tag{9.14}$$

Its interpretation in terms of (9.13) is the obvious one: to an element of D_0/nD_0, it associates its $\overline{v}$-coefficient in $\mathbf{Z}/n\mathbf{Z}$. (This is easily checked using Remark 7.2.)

9.15. End of the proof for k perfect. Note that (9.13) implies that $\operatorname{Hom}(N, \boldsymbol{\mu}_n)$ is a free $\mathbf{Z}/n\mathbf{Z}$-module. It follows — see also Exerc. 9.22 — that $\operatorname{Hom}(N, C) = \operatorname{Hom}(N, \boldsymbol{\mu}_n) \otimes C(-1)$, and that we have an exact sequence

$$0 \to \operatorname{Hom}(N, C) \to D/nD \otimes C(-1) \to C(-1) \to 0. \tag{9.16}$$

This exact sequence is compatible with the action of Γ_k. The group Γ_k acts on $\overline{V}$ with finite orbits, each orbit consisting of the elements lying above a given element v of V. The subgroup of Γ_k fixing an element $\overline{v}$ of $\overline{V}$ with image v in V is $\Gamma_{k(v)}$, and this describes D as

$$D = \oplus_{v \in V} \operatorname{ind}_{\Gamma_k}^{\Gamma_{k(v)}} \mathbf{Z}.$$

By the Shapiro-Faddeev Lemma, $H^{i-1}(k, D \otimes C(-1))$ may be identified with $\oplus_v H^{i-1}(k(v), C(-1))$, and under this identification the map onto $H^{i-1}(k, C(-1))$ induced by $D \otimes C(-1) \to C(-1)$ is given by the corestriction, which proves (9.7).

It remains to prove (9.8). If v, $\overline{v}$, and w are as above, we may apply Th. 6.1 and Remark 6.4 to the diagram (9.6). We thus obtain the commutativity of the diagram

$$\begin{array}{ccc} H^i(K, C) & \longrightarrow & H^{i-1}(k, \operatorname{Hom}(N, C)) \\ \downarrow & & \downarrow \\ H^i(\operatorname{Dec}_w, C) & \longrightarrow & H^{i-1}(k(v), \operatorname{Hom}(I_w, C)), \end{array}$$

which we may also rewrite as:

$$\begin{array}{ccc} H^i(K, C) & \xrightarrow{\ r\ } & \ker\Big(\oplus H^{i-1}(k(v), C(-1)) \to H^{i-1}(k, C(-1))\Big) \\ \downarrow & & \downarrow{\scriptstyle j_w} \\ H^i(\operatorname{Dec}_w, C) & \xrightarrow{\ r_v\ } & H^{i-1}(k(v), C(-1)). \end{array} \tag{9.17}$$

By what we have seen above (see (9.14)), j_w is the v-th component map. This proves (9.8) and completes the proof of Th. 9.2 when k is perfect.

9.18. GENERAL CASE. Let $k' = k^{p^{-\infty}}$ be the perfect closure of k. We have seen that Theorem 9.2 holds for k'; we are going to deduce it for k.

Note first that the places 'v' are the same for $k(t)$ and $k'(t)$: the canonical map $\mathbb{P}_{1/k'} \to \mathbb{P}_{1/k}$ is a homeomorphism. If we denote by v' the valuation of $k'(t)$ extending the valuation v of $k(t)$, then:

(1) the ramification index $e(v'/v)$ is equal to $q_v = [k(v) : k]_i$;
(2) the residue field $k'(v')$ of v' is $k'k(v)$; its degree over k' is $[k(v) : k]/q_v$.

Consider now the diagram

$$\begin{array}{ccccccc} H(k,C) & \longrightarrow & H(k(t),C) & \longrightarrow & \oplus_v H(k(v),C(-1)) & \longrightarrow & H(k,C(-1)) \\ \downarrow{\scriptstyle\alpha} & & \downarrow{\scriptstyle\beta} & & \downarrow{\scriptstyle\gamma} & & \downarrow{\scriptstyle\delta} \\ H(k',C) & \longrightarrow & H(k'(t),C) & \longrightarrow & \oplus_v H(k'(v),C(-1)) & \longrightarrow & H(k',C(-1)), \end{array} \tag{9.19}$$

where α, β, δ are the natural isomorphisms associated with the purely inseparable extensions k'/k, $k'(t)/k(t)$, and k'/k, and $\gamma = \oplus_v \gamma_v$ where γ_v is the natural isomorphism $H(k(v),C(-1)) \to H(k'(v),C(-1))$ multiplied by q_v. All the vertical maps are isomorphisms (because q_v is prime to the order of C).

To prove Th. 9.2 for k, it is sufficient to prove that Diagram (9.19) commutes, since Th. 9.2 holds for k'. The commutativity of the $\alpha\beta$ square is clear. That of the $\beta\gamma$ square follows from the commutativity of 8.4. As for commutativity of the $\gamma\delta$ square, it follows from:

LEMMA 9.20. *Let k_1 be a finite extension of k; let $q = [k_1 : k]_i$ and let $k_1' = k'k_1$. If A is any Γ_k-module, the diagram*

$$\begin{array}{ccc} H(k_1,A) & \xrightarrow{\text{Cor}} & H(k,A) \\ \downarrow{\scriptstyle\varphi_1} & & \downarrow{\scriptstyle\varphi} \\ H(k_1',A) & \xrightarrow{\text{Cor}} & H(k',A) \end{array}$$

is commutative, where φ is the natural isomorphism $H(k,A) \to H(k',A)$ and φ_1 is the natural isomorphism $H(k_1,A) \to H(k_1',A)$ multiplied by q.

PROOF. This is easily checked when k_1/k is separable (i.e., when $q = 1$) or is purely inseparable (i.e., when $k_1' = k'$). The general case follows by dévissage. □

This completes the proof of Theorem 9.2.

EXAMPLE 9.21 (the Brauer group, see [Fa 51]). Let $C = \boldsymbol{\mu}_n$ for n not divisible by the characteristic, so that $C(-1) = \mathbf{Z}/n\mathbf{Z}$. We have $H^2(K,C) = \mathrm{Br}_n(K)$,[a] the kernel of multiplication by n on the Brauer group $\mathrm{Br}(K)$. The exact sequence in 9.2 can be written:

$$0 \to \mathrm{Br}_n(k) \to \mathrm{Br}_n(K) \to \oplus_{v \in V} H^1(k(v), \mathbf{Z}/n\mathbf{Z}) \to H^1(k, \mathbf{Z}/n\mathbf{Z}) \to 0.$$

EXERCISE 9.22. Let K be a field containing a primitive n-th root of unity.

(1) Show that $\mathrm{Hom}(\Gamma_K, \boldsymbol{\mu}_n)$ is a projective $\mathbf{Z}/n\mathbf{Z}$-module (and even a free module if n is a power of a prime).

[a] There are two natural identifications of $H^2(K,\boldsymbol{\mu}_n)$ with $\mathrm{Br}_n(K)$, differing by a sign, see [KMRT 98, pp. 396, 397]. Which one is chosen here makes no difference to the statement.

(2) Let C be a Γ_K-module with $nC = 0$. Prove that the natural map

$$\mathrm{Hom}(\Gamma_K, \boldsymbol{\mu}_n) \otimes C(-1) \to \mathrm{Hom}(\Gamma_K, C)$$

is an isomorphism.

[Hint: For (1), use the following criterion for a $\mathbf{Z}/n\mathbf{Z}$-module E to be projective (or injective): for every m dividing n and every $x \in E$ with $mx = 0$, there exists $y \in E$ with $(n/m) \cdot y = x$. One applies this with $E = K^*/K^{*n}$. Note that "projective" is equivalent to "free" when n is a power of a prime. As for (2), it follows from (1).]

EXERCISE 9.23 (value of a cohomology class at a rational point). Fix some $\alpha \in H^i(k(t), C)$ and $x \in \mathbb{P}_1(k)$, $x \neq \infty$. Assume that α is unramified at infinity and at x. Prove that

$$\alpha(x) - \alpha(\infty) = \sum_{\substack{v \in V \\ v \neq x}} \mathrm{Cor}_k^{k(v)} \left[(x - t_v) \cdot r_v(\alpha)\right].$$

(In the expression $(x - t_v)$, x is viewed as an element of k, and t_v denotes the image of $t \in k[t]$ in $k(v)$. Then $(x - t_v)$ is the corresponding class in $H^1(k(v), \boldsymbol{\mu}_n)$ and $(x - t_v) \cdot r_v(\alpha)$ is its cup product with $r_v(\alpha) \in H^{i-1}(k(v), C(-1))$.)

9.24. GENERALIZATION TO ALGEBRAIC CURVES. Part of what has been done above for the projective line can be extended to an arbitrary connected smooth projective curve X over k. We limit ourselves to stating the results. Let V be the set of closed points of X. We write K for the function field of X over k. An element $v \in V$ can be identified with a discrete valuation of K which is trivial on k; we write $k(v)$ for the corresponding residue field, which is a finite extension of k.

It is still true that every element in $H^i(K, C)$ has only a finite number of poles, hence the sequence occurring in 9.2 is still defined. The residue formula also holds; it can be proved by writing the curve as a (ramified) covering of the projective line and using a corestriction argument.

Let us assume that X has a rational k-point (or, more generally, a closed point whose degree is prime to n, where $nC = 0$). In that case, the map $H^i(k, C) \to H^i(K, C)$ is injective, and one has an exact sequence

$$(9.25) \qquad \cdots \to H^{i-1}(k, J_n \otimes C(-1)) \to H^i(K, C)/H^i(k, C) \xrightarrow{r}$$
$$\xrightarrow{r} \oplus^0_{v \in V} H^{i-1}(k(v), C(-1)) \to H^i(k, J_n \otimes C(-1)) \to \cdots,$$

where:

- J_n is the group of n-division points[b] of the Jacobian of X; it is a free $\mathbf{Z}/n\mathbf{Z}$-module of rank $2g$, where g is the genus of X;
- $\oplus^0_{v \in V} H^{i-1}(k(v), C(-1))$ is the subgroup of $\oplus_{v \in V} H^{i-1}(k(v), C(-1))$ made up of the elements whose sum of corestrictions in $H^{i-1}(k, C(-1))$ is 0;
- the map r is given by the residues relative to the elements v of V.

Note that the kernel of r is usually $\neq 0$. If, for instance, k is algebraically closed, $C = \boldsymbol{\mu}_n$, and $i = 1$, this kernel is isomorphic to J_n.

10. Galois cohomology of $k(t_1, \ldots, t_n)$

As before, C is a finite Γ_k-module with order prime to the characteristic.

---

[b]There is a canonical isomorphism $\wedge^2 J_n \cong \boldsymbol{\mu}_n$. Hence the Γ_k-module $J_n \otimes C(-1)$ may also be written as $J_n^* \otimes C$ or as $\mathrm{Hom}(J_n, C)$.

THEOREM 10.1. *Let $n \geq 1$ and $K = k(t_1, \ldots, t_n)$ where the t_i are independent indeterminates. Then*

1. *The natural map $H^i(k, C) \to H^i(K, C)$ is injective.*
2. *Let α be an element of $H^i(K, C)$, whose residues are 0 at all discrete valuations of K/k which correspond to irreducible hypersurfaces of the n-dimensional affine space Aff^n. Then α is constant, i.e., belongs to $H^i(k, C)$.*

PROOF. Both parts are proved by induction on n, and part (1) is immediate from the case $n = 1$, cf. Theorem 9.2.

Set $K' = k(t_1, \ldots, t_{n-1})$, the function field of Aff^{n-1}, and consider the projection map $\pi \colon \mathrm{Aff}^n \to \mathrm{Aff}^{n-1}$ given by forgetting the last coordinate. The generic fiber is an affine line $\mathrm{Aff}^1_{/K'}$ over K' whose coordinate ring is $K'[t_n]$; its function field is $K = K'(t_n)$.

The irreducible divisors on Aff^n are of two kinds:

(a) The vertical ones, i.e., those which are the inverse image under π of an irreducible divisor of Aff^{n-1}; they correspond to irreducible polynomials $P(t_1, \ldots, t_n)$ which are independent of t_n.

(b) The non-vertical ones, which correspond to the closed points of $\mathrm{Aff}^1_{/K'}$.

Since the residues of α are zero for all places of type (b), 9.3 shows that α belongs to $H^i(K', C)$.

Let us now show that α is constant (when viewed as an element of $H^i(K', C)$). Let D' be an irreducible divisor of Aff^{n-1} with function field F'. The inverse image of D' in Aff^n is a divisor D whose function field is $F = F'(t_n)$. By 8.2.1 applied to K' and K (instead of K and K') with $e = 1$, the diagram

$$\begin{array}{ccc} H^i(K', C) & \xrightarrow{r_{D'}} & H^{i-1}(F', C(-1)) \\ \downarrow & & \downarrow \\ H^i(K, C) & \xrightarrow{r_D} & H^{i-1}(F, C(-1)) \end{array}$$

commutes, where the horizontal arrows are the residue maps and the vertical arrows are injections (9.3). Since $\alpha \in H^i(K, C)$ has residue zero at D, it also has residue zero at D' when viewed as an element of $H^i(K', C)$. This shows that all the residues of α are 0; by the induction assumption applied to $K' = k(t_1, \ldots, t_{n-1})$, this implies that α is constant. $\square$

REMARK 10.2. The conditions of 10.1.2 refer only to the behavior of α along the divisors of Aff^n. Hence, if α "behaves well" in codimension 1, it does so everywhere. One should compare this "purity" statement with the analogous (but deeper) result of Bloch, Ogus, and Gabber for étale cohomology (see e.g. [CHK 97]).

CHAPTER IV

Specialization properties of cohomological invariants

11. Compatibility with good reduction

Let G be a smooth algebraic group over k_0, and let C be a finite Γ_{k_0}-module of order not divisible by the characteristic.[a]

We consider an extension K of k_0, with a discrete valuation v; we call its valuation ring R and its residue field k. We assume that R contains k_0, so that R and K are k_0-algebras; k is also a k_0-algebra via the composition $k_0 \to R \to k$. Until 11.7, we assume also that K is *complete.* (This assumption could be replaced by "henselian".)

Let T_k be a k-G-torsor. It is well known that there is an R-G-torsor T_R which has "special fiber" T_k and that T_R is unique up to isomorphism, cf. [SGA 3, p. 401, Prop. 8.1]. Since any R-G-torsor T_R gives by base extension a K-G-torsor, this defines a map $i\colon H^1(k,G) \to H^1(K,G)$.

The following theorem and its proof are due to M. Rost (letter to Serre dated 2 May 1994). We reproduce them here with his permission:

COMPATIBILITY THEOREM 11.1. *For $a \in \mathrm{Inv}_{k_0}(G,C)$, the diagram*

$$\begin{array}{ccc} H^1(k,G) & \xrightarrow{\ i\ } & H^1(K,G) \\ {\scriptstyle a_k}\downarrow & & \downarrow{\scriptstyle a_K} \\ H(k,C) & \xrightarrow{\ j\ } & H(K,C) \end{array}$$

commutes.

The map j is the natural embedding defined in 7.6 and 7.7.

PROOF OF THEOREM 11.1. We have to show that the equation

$$j(a_k(x)) = a_K(i(x)) \tag{11.2}$$

holds for all $x \in H^1(k,G)$.

We need a few lemmas:

LEMMA 11.3. *$H^1(k,G)$ is the inductive limit of the $H^1(k_1,G)$ where k_1 runs through the finitely generated extensions of k_0 contained in k.*

This is clear.

LEMMA 11.4. *If K_1/K is purely inseparable, the map $H(K,C) \to H(K_1,C)$ is bijective.*

[a] As Rost has pointed out to us, no hypothesis on the Γ_{k_0}-module C is needed until 11.7.

The map $\Gamma_K \to \Gamma_{K_1}$ is an isomorphism, hence the lemma.

Consider now a subfield k_1 of k containing k_0. We say that k_1 is *K-liftable* if there is a k_0-morphism $k_1 \to R$ such that its composite with $R \to k$ is equal to the injection $k_1 \to k$. In other words, it is a lifting (in the usual sense) of k_1 in R which extends the given lifting $k_0 \to R$.

LEMMA 11.5. *Suppose k_1 is K-liftable, and $x \in H^1(k, G)$ is the image of an element x_1 of $H^1(k_1, G)$* (i.e., x "comes from k_1"). *Then* (11.2) *holds for x.*

PROOF. Choose a k_1-G-torsor T_1 whose class in $H^1(k_1, G)$ is x_1. A lifting $k_1 \to R$ transforms T_1 by base change into an R-G-torsor. This is the T_R mentioned at the beginning of the section. The embedding $R \to K$ transforms T_R into T_K. The class of T_K is $i(x)$. But it is also the image of x_1 via $k_1 \to K$. Hence $a_K(i(x))$ is equal to $j_1(a_{k_1}(x_1))$, where a_{k_1} is relative to k_1 and j_1 comes from the lifting $k_1 \to R \to K$.

The diagram

$$\begin{array}{ccc} H(k_1, C) & \xrightarrow{j_1} & H(K, C) \\ \downarrow & & \| \\ H(k, C) & \xrightarrow{j} & H(K, C) \end{array}$$

commutes, where the left-hand arrow comes from the embedding $k_1 \to k$. Because a is functorial, we have $j_1(a_{k_1}(x_1)) = j(a_k(x))$, which proves the lemma. $\square$

LEMMA 11.6. *Let k_1 be a finitely generated extension of k_0 contained in k. Then there exists a finite purely inseparable extension K_1 of K such that k_1 is K_1-liftable.*

(The statement of the lemma makes sense because the discrete valuation of K extends in a unique way to a discrete valuation of K_1, which is also complete.)

PROOF. *Case I*: *k_1 is a finite purely inseparable extension of k_0.* The two extensions k_1/k_0 and K/k_0 can be embedded in some extension K_2. Define K_1 to be the subfield of K_2 generated by k_1 and K. It has the required properties.

Case II: *k_1 is a finite extension of k_0.* Let k' be the largest separable extension of k_0 contained in k_1. The extension k_1/k' is purely inseparable. By Hensel's Lemma, k' is K-liftable. One then applies Case I, with k_0 replaced by k'.

Case III: *The general case.* Let $t_1, \ldots, t_d$ be a transcendence basis of k_1/k_0, and let $k_0(\mathbf{t})$ be the extension of k_0 generated by $t_1, \ldots, t_d$. The field k_1 is a finite extension of $k_0(\mathbf{t})$. The field $k_0(\mathbf{t})$ is K-liftable, as one sees by choosing R-representatives of the t_i. One then applies Case II, with k_0 replaced by $k_0(\mathbf{t})$. $\square$

END OF PROOF OF THEOREM 11.1. Let x be an element of $H^1(k, G)$. By 11.3, we may choose k_1 finitely generated over k_0 such that x comes from k_1. By 11.5 and 11.6 we may choose K_1/K as in 11.6 such that (11.2) holds after replacing K by K_1. This means that the two sides of (11.2) have the same image in $H(K_1, C)$. By the injectivity property 11.4, these two sides are equal. $\square$

11.7. Suppose again that we have a discrete valuation ring R containing a field k_0, with quotient field K and residue field k, and corresponding to a valuation v. That is, R is as above except that we do not assume that R is complete. Any R-G-torsor T_R gives torsors T_K and T_k.

THEOREM. *For $a \in \mathrm{Inv}_{k_0}(G,C)$ and $\alpha = a(T_K)$, we have:*

(1) *The residue of α at v is* 0.

(2) *The value of α at v (in the sense of* 7.13) *is $a(T_k)$.*

PROOF. We may assume that R is complete, in which case this follows from Th. 11.1. □

APPLICATION 11.8. Suppose we have a G-torsor T over X where X is a normal irreducible k-variety (and k is an extension of k_0). Let K be the function field of X, and write T_K for the torsor over K obtained from T. The following is a direct consequence of 11.7: For every invariant a, $a(T_K)$ is "unramified at all irreducible divisors of X", i.e., it is unramified (7.13) at the valuation defined by such an irreducible divisor.

12. Specialization properties

12.1. Here and in the following theorem, let R be a noetherian domain which contains k_0. Let K be the quotient field of R, which is an extension of k_0. Let $\mathfrak{m}$ be a maximal ideal of R with residue field k. An R-G-torsor T_R defines torsors T_K over K and T_k over k.

SPECIALIZATION THEOREM 12.2. *Suppose that $R_\mathfrak{m}$ is a regular local ring. Let T_R be an R-G-torsor, and denote by T_k and T_K the k-G- and K-G-torsors defined by T_R. Let a be an invariant in* $\mathrm{Inv}(G,C)$. *If $a(T_K)=0$, then $a(T_k)=0$.*

PROOF. We may replace R with $R_\mathfrak{m}$ and so assume that R is local. We use induction on the dimension of R.

If R has dimension 1, then it is a discrete valuation ring. By Theorem 11.7.2, $a(T_k)$ is the value of $a(T_K)$, which is 0.

If $n = \dim R$ is > 1, $\mathfrak{m}$ is generated by n elements $t_1, \ldots, t_n$. Let $R' = R/t_1R$. This ring has dimension $n-1$ and is also regular; let K' be its quotient field. By the dimension 1 case, we have $a(T_{K'}) = 0$. By the induction hypothesis (applied to R' and K'), we get $a(T_k) = 0$. □

12.3. COHOMOLOGICAL INVARIANTS AND VERSAL TORSORS. Let $P \in H^1(K,G)$ be a versal element in the sense of 5.1. Such an element P detects the equality of cohomological invariants, i.e.:

THEOREM. *Let a, b be two invariants in* $\mathrm{Inv}_{k_0}(G,C)$. *If $a(P) = b(P)$ in $H(K,C)$, then $a=b$.*

PROOF. By replacing a and b with $a-b$ and 0 respectively, we may assume that $b=0$.

We have to show that $a(T)=0$ for every k/k_0 and every k-G-torsor T. Suppose first that k is infinite. By the versal property, T is isomorphic to Q_x for some $x \in X(k)$ with the notation of 5.1. Let R be the local ring of $X_{/k}$ at x. Its residue field is k. Let K_x be its field of fractions, i.e., the function field of $X_{/k}$. Since X is smooth, the ring R is regular; let Q_R be the pullback of Q by the map $\mathrm{Spec}(R) \to X$. The fiber of Q_R at x (resp. at the generic point η) is T (resp. Q_η). We have $a(Q_\eta)=0$ because Q_η is obtained from P by the base change $K \to K_x$. Hence, by 12.2, we have $a(T)=0$.

Now suppose that k is finite. The torsor T defines an element $T_t \in H^1(k(t), G)$ by the base extension $k \to k(t)$, and by the case where k is infinite, we have $a(T_t) = 0$. So $a(T)$ is in the kernel of the map $H(k, C) \to H(k(t), C)$, which is zero by 9.2. Hence $a(T) = 0$. □

COROLLARY 12.4. *Assume k_0 is algebraically closed. Then* $\mathrm{Inv}^i(G, C) = 0$ *for every $i > \mathrm{ed}(G)$.*

PROOF. By the definition of $\mathrm{ed}(G)$, there exists a versal G-torsor P defined over a field K with transcendence degree $\mathrm{ed}(G)$ over k_0. By a well-known result of Grothendieck-Tate (see e.g. [Se 65, §II.4.2]) one has $H^i(K, C) = 0$, hence $\mathrm{Inv}^i(G, C)$ is 0 by Th. 12.3. □

Hence any non-zero invariant gives a lower bound for $\mathrm{ed}(G)$. For instance, assume the characteristic is $\neq 2$, and take $G = \mathbf{O}(n)$; the non-vanishing of the Stiefel-Whitney invariant w_n (see §17) implies that $\mathrm{ed}(G) \geq n$; since it is easy to construct a versal torsor with $\mathrm{tr.deg}(K) = n$, this shows that $\mathrm{ed}(\mathbf{O}(n)) = n$. A similar argument gives the values of $\mathrm{ed}(\mathbf{SO}(n))$ and $\mathrm{ed}(G_2)$ mentioned in 5.7, see [Re 00]. In the case of the symmetric group S_n, this method gives only the bound $\mathrm{ed}(S_n) \geq [n/2]$, which is probably not sharp for large n's.

REMARK 12.5. Th. 12.3 shows that the map $\mathrm{Inv}(G, C) \to H(K, C)$ given by $a \mapsto a(P)$ is an injection. We may then view $\mathrm{Inv}(G, C)$ as a *subgroup* of $H(K, C)$. By Th. 11.7, the elements of that subgroup are "unramified over X" in the sense that they have residue 0 for all the discrete valuations of K defined by the irreducible divisors of X. There are important cases where the converse is true, i.e., where $\mathrm{Inv}(G, C)$ is *equal* to the subgroup of $H(K, C)$ made up of the elements which are unramified over X, cf. B. Totaro's letter reproduced in Appendix C.

CHAPTER V

Restriction and corestriction of invariants

13. Restriction of invariants

If $f : H \to G$ is a homomorphism of algebraic groups (over k_0), any H-torsor T_1 (over an extension k/k_0) defines a G-torsor $T = T_1.G$. Hence any invariant $a \in \mathrm{Inv}(G, \mathsf{H})$ defines an invariant $f^*a \in \mathrm{Inv}(H, \mathsf{H})$, where H is any functor as in §1; one has $f^*a(T_1) = a(T)$. When H is a subgroup of G and f is the injection of H in G, the map

$$f^* : \mathrm{Inv}(G, \mathsf{H}) \to \mathrm{Inv}(H, \mathsf{H})$$

is called *restriction* (and is denoted by Res_G^H or simply by Res when there is no ambiguity regarding H and G).

PROPOSITION 13.1. *Let f, $f' : H \to G$ be two homomorphisms. Assume they are conjugate by an element x in $G(k_0)$* (i.e., if we write i_x for the inner automorphism defined by x, we have $f' = i_x \circ f$). *Then f^* is equal to $(f')^*$.*

PROOF. Let $a \in \mathrm{Inv}(G, \mathsf{H})$, and let $b = f^*a$ and $b' = (f')^*a$ be the corresponding elements of $\mathrm{Inv}(H, \mathsf{H})$. We have to show that, if T_1 is any H-torsor over an extension k of k_0, one has $b(T_1) = b'(T_1)$. Let T and T' be the G-torsors deduced from T_1 via f, f'. The torsor T is the quotient of $T_1 \times G$ by H acting by

$$h \cdot (t, g) = (th^{-1}, f(h)g),$$

while the action of G on the right comes from

$$(t, g) \cdot g' = (t, gg').$$

The same is true for T', with f replaced by f'. One then defines a map $T_1 \times G \to T_1 \times G$ by

$$(t, g) \mapsto (t, xg).$$

This defines (by passage to the quotients) an isomorphism of T onto T'. This implies that $a(T) = a(T')$. We then have

$$b(T_1) = a(T) = a(T') = b'(T_1). \qquad \square$$

13.2. Suppose H is a subgroup of G and write N for its normalizer in G. Every element of $N(k_0)$ defines an automorphism of H, hence an automorphism of $\mathrm{Inv}(H, \mathsf{H})$.

PROPOSITION. (1) *The action of $N(k_0)$ on $\mathrm{Inv}(H, \mathsf{H})$ factors through the quotient $N(k_0)/H(k_0)$.*

(2) *If $a \in \mathrm{Inv}(H, \mathsf{H})$ is the restriction of an element of $\mathrm{Inv}(G, \mathsf{H})$, then a is fixed by $N(k_0)/H(k_0)$.*

PROOF. This follows from Prop. 13.1. $\square$

REMARK 13.3. One may ask the following questions:

(a) When is Res: $\mathrm{Inv}(G, \mathsf{H}) \to \mathrm{Inv}(H, \mathsf{H})$ injective?
(b) What is the image of Res? In particular, when is it equal to the set of elements fixed by $N(k_0)/H(k_0)$?

When G is finite, we give in §15 some partial answers to these questions.

14. Corestriction of invariants

(The material in §14 and §15 will only be used in §34.)

14.1. In this section, we assume that G is a finite group (viewed as a "constant" group scheme of dimension 0) and H is a subgroup of G. We also assume that the functor H is the functor $k \mapsto H(k, C)$ associated with some Galois module C over k_0.

14.2. Before defining the corestriction, it is convenient to make some definitions. If k is an extension of k_0 and K is an étale k-algebra with $K = \prod_{i=1}^{n} k_i$ (with the k_i finite extensions of k), we put

$$H(K, C) = \prod_{i=1}^{n} H(k_i, C) \quad \text{and} \quad H^1(K, G) = \prod_{i=1}^{n} H^1(k_i, G).$$

We write Cor: $H(K, C) \to H(k, C)$ for the sum of the corestrictions $H(k_i, C) \to H(k, C)$. If $Y = \mathrm{Spec}(K)$ is the étale k-scheme associated with K, we also write $H(Y, C)$ for $H(K, C)$, and Cor^Y or Cor_k^K for Cor: $H(K, C) \to H(k, C)$.

(Note that $H(Y, C)$ is the étale cohomology of $Y = \coprod \mathrm{Spec}(k_i)$, with coefficients in the sheaf defined by C.)

If a is a cohomological invariant in $\mathrm{Inv}(G, C)$, and if $x = (x_1, \ldots, x_n)$ is an element of $H^1(K, G)$, then $a(x) \in H(K, C)$ is defined as the tuple $(a(x_1), \ldots, a(x_n))$.

DEFINITION 14.3. Let a be in $\mathrm{Inv}(H, C)$, and let T be a G-torsor over k. The action of H on T makes T into an H-torsor T_H over T/H, and T/H is a finite étale k-scheme, hence of the form $\mathrm{Spec}(K)$, where K is an étale k-algebra, cf. 2.1. By the above, $a(T_H)$ is a well-defined element of $H(K, C)$. Put

$$b(T) = \mathrm{Cor}(a(T_H)) \quad \text{in } H(k, C),$$

where Cor: $H(K, C) \to H(k, C)$ is as in 14.2 above. It is clear that $T \mapsto b(T)$ is an element of $\mathrm{Inv}(G, C)$, which we call the *corestriction of a* and denote by $\mathrm{Cor}_G^H(a)$ or simply by $\mathrm{Cor}(a)$. This defines a homomorphism

$$\mathrm{Cor}\colon \mathrm{Inv}(H, C) \to \mathrm{Inv}(G, C).$$

We have the usual formula:

PROPOSITION 14.4. *$\mathrm{Cor} \circ \mathrm{Res} = n$, where $n = (G : H)$.*

PROOF. Let a be an element of $\mathrm{Inv}(G, C)$. Let a_1 be its restriction to H and put $b = \mathrm{Cor}(a_1)$. We have to prove that $b = n \cdot a$, i.e., that

$$b(T) = n \cdot a(T)$$

for every G-torsor T over every extension k of k_0. As in Def. 14.3, let T_H be the H-torsor over T/H defined by the action of H on T, and let us write T/H as

Spec(K), where K is an étale k-algebra. If $T_H.G$ denotes the G-torsor deduced from T_H by the embedding $H \to G$, we have

$$a_1(T_H) = a(T_H.G) \text{ in } H(K,C) \quad \text{and} \quad b(T) = \mathrm{Cor}_k^K(a_1(T_H)) \text{ in } H(k,C).$$

There is a commutative diagram of G-torsors:

$$\begin{array}{ccc} T_H.G & \longrightarrow & T \\ \downarrow & & \downarrow \\ \mathrm{Spec}(K) = T/H & \longrightarrow & T/G = \mathrm{Spec}(k), \end{array}$$

where the top map is given by $(t,g) \mapsto tg$. This diagram shows that the G-torsor $T_H.G$ is isomorphic to the pullback of T by $T/H \to T/G = \mathrm{Spec}(k)$; hence $a(T_H.G)$ is the image of $a(T)$ by the natural map $H(k,C) \to H(K,C)$. Since $[K:k] = n$, this shows that $\mathrm{Cor}(a(T_H.G)) = n\cdot a(T)$, i.e., that $b(T) = n\cdot a(T)$. □

14.5. THE DOUBLE COSET FORMULA (À LA MACKEY). Let H and H' be two subgroups of G. We wish to compute the map

$$\mathrm{Res}_G^{H'} \circ \mathrm{Cor}_G^H : \mathrm{Inv}(H,C) \to \mathrm{Inv}(H',C).$$

To do so, choose a set S of representatives of $H\backslash G/H'$. For each $s \in S$, one defines

$$H_s = sH's^{-1} \cap H \quad \text{and} \quad H'_s = s^{-1}Hs \cap H'.$$

One has $s^{-1}H_s s = H'_s$. Define

$$(14.6) \qquad f_s : \mathrm{Inv}(H,C) \xrightarrow{\mathrm{Res}_H^{H_s}} \mathrm{Inv}(H_s,C) \xrightarrow{i_s^*} \mathrm{Inv}(H'_s,C) \xrightarrow{\mathrm{Cor}_{H'}^{H'_s}} \mathrm{Inv}(H',C),$$

where i_s^* is the isomorphism associated with the conjugation $i_s : H'_s \to H_s$, cf. Prop. 13.1. (The right arrow makes sense because $H'_s \subset H'$.)

REMARK 14.7. Note that, when s is replaced by hsh', with $h \in H$ and $h' \in H'$, then H_s is replaced by $h\,H_s\,h^{-1}$ and H'_s by $(h')^{-1}\,H'_s\,h'$. This does not change f_s, because of Prop. 13.1.

Note also that $i_s^* \circ \mathrm{Res}_H^{Hs}$ is none other than the restriction map

$$\phi_s^* : \mathrm{Inv}(H,C) \to \mathrm{Inv}(H'_s,C)$$

corresponding to the embedding $\phi_s : H'_s \to H$ given by $x \mapsto sxs^{-1}$.

PROPOSITION 14.8. $\mathrm{Res}_G^{H'} \circ \mathrm{Cor}_G^H = \sum_{s\in S} f_s$ *for* f_s *as in* (14.6).

PROOF. Let a be in $\mathrm{Inv}(H,C)$. Write a_s for the image of a in $\mathrm{Inv}(H_s,C)$ by restriction; write a'_s for the image of a_s in $\mathrm{Inv}(H'_s,C)$ by conjugation, and put $c_s = \mathrm{Cor}_{H'}^{H'_s}(a'_s) = f_s(a)$, which belongs to $\mathrm{Inv}(H',C)$. Let $b = \mathrm{Cor}_G^H(a)$ and $c = \mathrm{Res}_G^{H'}(b)$. We have to show that $c = \sum c_s$, i.e., that, if T' is an H'-torsor over an extension k/k_0, we have $c(T') = \sum c_s(T')$.

Let $T'.G$ be the G-torsor deduced from T' by $H' \to G$; the natural action of H on $T'.G$ makes $T'.G$ into an H-torsor T over T/H; we denote the étale k-scheme T/H by Y. We have $a(T) \in H(Y,C)$, and the definition of c shows that

$$c(T') = \mathrm{Cor}^Y(a(T)).$$

If $s \in S$, let $G_s = H's^{-1}H$ be the double coset of G containing s^{-1}. We have $G = \coprod_{s\in S} G_s$ (disjoint union). This decomposition of G gives a corresponding decomposition of $T'.G$ as $T'.G = \coprod T'.H's^{-1}H$. Hence the torsor T decomposes as the disjoint union of the torsors $T_s = T'.H's^{-1}H$; let $Y_s = T_s/H$ be the base of T_s. It is not difficult to see that T_s is deduced from T' by first restricting the group action from H' to H'_s, and then taking the corresponding H-torsor via the embedding $\phi_s\colon H'_s \to H$ of Remark 14.7. This shows that $a(T_s) = a'_s(T')$.

We have $Y = \coprod Y_s$, hence $H(Y,C) = \prod H(Y_s,C)$. By what we have just seen, the components of $a(T)$ are the $a'_s(T')$. The corestriction of $a(T)$ is the sum of the corestrictions of its components. This gives

$$\mathrm{Cor}^Y(a(T)) = \sum \mathrm{Cor}^{Y_s}(a'_s(T')),$$

i.e., $c(T') = \sum c_s(T')$. □

REMARK 14.9. The decomposition of T as $\coprod T_s$ in the proof of Prop. 14.8 is a general fact about torsors which is true over any base (not necessarily the spectrum of a field).

15. Stability

DEFINITION 15.1. We keep the notation of 13.2 and 14.1. Let a be an element of $\mathrm{Inv}(H,C)$. As in [CE 56, XII.9], we say that a is *stable* (with respect to G) if it has the following property:

(15.2) *For every $H' \subset H$ and every $s \in G$ such that $sH's^{-1} \subset H$, the two maps i, $j\colon H' \to H$ defined by*

$$i(x) = x \qquad \textit{and} \qquad j(x) = sxs^{-1}$$

are such that $i^*(a) = j^*(a)$ in $\mathrm{Inv}(H',C)$.

Note the special case of (15.2) where $H' = H$. In that case, s belongs to the normalizer N of H in G, and the condition $i^*(a) = j^*(a)$ means that a is fixed under the action of N/H on $\mathrm{Inv}(H,C)$ defined in 13.2.

PROPOSITION 15.3. (1) *Every element of the image of the restriction map $\mathrm{Inv}(G,C) \to \mathrm{Inv}(H,C)$ is stable.*

(2) *Let a be a stable element of $\mathrm{Inv}(H,C)$. Then*

$$\mathrm{Res}(\mathrm{Cor}(a)) = n\cdot a \quad \textit{where } n = (G:H).$$

PROOF. Assertion (1) follows from 13.1. Assertion (2) follows from the double coset formula 14.8 applied to $H' = H$. Indeed, $\mathrm{Res}(\mathrm{Cor}(a))$ is the sum of the $f_s(a)$, and the stability of a implies that $f_s(a) = n_s\cdot a$ where $n_s = (H:H_s)$; one then uses the fact that n is the sum of the n_s, as one sees by writing G/H as a disjoint union of H-orbits indexed by S. □

The main consequence of the proposition is:

COROLLARY 15.4. *If $(G:H)$ and $|C|$ are relatively prime, then*

$$\mathrm{Res}\colon \mathrm{Inv}(G,C) \to \mathrm{Inv}(H,C)$$

is injective, and its image is the subgroup of $\mathrm{Inv}(H,C)$ made up of the stable elements.

This applies in particular when H is a p-Sylow subgroup of G, and C is a p-group.

REMARK 15.5. The stability condition (15.2) involves all the subgroups H' of H. There are special cases where one can dispense with most of them. For instance, let N be a subgroup of G which contains H and which has "strong control" of the fusion of H in G, in the following sense:

(Fus) *For every subgroup H' of H and every $s \in G$ with $sH's^{-1} \subset H$, there exists $t \in N$ with $txt^{-1} = sxs^{-1}$ for every $x \in H'$.*

PROPOSITION 15.6. *Suppose that N normalizes H and has property* (Fus) *above. Then an element of* $\mathrm{Inv}(H, C)$ *is stable if and only if it is fixed under the action of N/H.*

This follows from Prop. 13.1.

EXAMPLE 15.7. If H is a p-Sylow of G which is *abelian* and N is its normalizer, a well-known theorem of Burnside says that (Fus) is true. Let us recall the proof. If s and H' are as in (Fus), let G' be the centralizer of H' in G. Since H is abelian and contains H', we have $H \subset G'$. Since $s^{-1}Hs \supset H'$, we also have $s^{-1}Hs \subset G'$. By the conjugacy of p-Sylows in G', there exists $c \in G'$ with $cHc^{-1} = s^{-1}Hs$. The element $t = sc$ has the required properties.

Hence, if C is a p-group, $\mathrm{Inv}(G, C)$ may be identified with the subgroup of $\mathrm{Inv}(H, C)$ fixed under the action of N/H.

EXAMPLE 15.8. Take:

$$\begin{aligned} G &= S_n = \text{group of permutations of } \{1, \dots, n\}, \\ H &= S_{n-1} = \text{stabilizer of } n, \\ N &= H = S_{n-1}. \end{aligned}$$

Then (Fus) is true. Indeed, if s and H' are as in (Fus), take $t = s$ if $s(n) = n$. If not, put $m = s^{-1}(n)$, and let $c \in G$ be the transposition $(m\,n)$. Since $sH's^{-1}$ is contained in H, it fixes n, hence H' fixes both m and n. This implies that c centralizes H'. One then puts $t = sc$. One has $t(n) = s(m) = n$, hence t belongs to H and it is clear that $txt^{-1} = sxs^{-1}$ for every $x \in H'$.

Hence every element of $\mathrm{Inv}(S_{n-1}, C)$ is stable. In particular, if n is prime to $|C|$, *the restriction map* $\mathrm{Inv}(S_n, C) \to \mathrm{Inv}(S_{n-1}, C)$ *is an isomorphism.* We shall prove more precise results later (25.14.2 and 31.23). Indeed, we shall see that the subgroup $\mathrm{Inv}^{\mathrm{norm}}(S_n, C)$ of $\mathrm{Inv}(S_n, C)$ is 0 if $|C|$ is odd, so that the interesting case is the one where n is odd and $|C|$ is a power of 2, in which case the restriction map $\mathrm{Inv}(S_n, C) \to \mathrm{Inv}(S_{n-1}, C)$ is an isomorphism.

PROPOSITION 15.9. *Suppose H is a p-Sylow of G, and C is killed by a power of p. Let a be an element of* $\mathrm{Inv}(H, C)$. *Assume*:

(1) *a is invariant under the automorphisms of H coming from the normalizer N of H in G;*
(2) $\mathrm{Res}^{H'}_{H}(a) = 0$ *for every proper subgroup H' of H.*

Then there is a unique element a_G of $\mathrm{Inv}(G, C)$ *such that* $\mathrm{Res}^{H}_{G}(a_G) = a$, *namely* $a_G = \mathrm{Cor}^{H}_{G}(a)/(N : H)$.

PROOF. The uniqueness is obvious, so it suffices to prove existence. Define b as $\mathrm{Cor}(a)$, and compute $\mathrm{Res}(b)$ by the double coset formula (with $H' = H$). The elements $s \in S$ which are not in N give 0 because of (2); those in N give a because of (1). Hence the restriction of b is a, multiplied by $(N : H)$ which is prime to p. □

CHAPTER VI

Cohomological invariants of O_n, SO_n, ...

In this chapter, the characteristic is assumed to be $\neq 2$.

16. Cohomological invariants of $(2,\ldots,2)$ groups

In this section we determine $\mathrm{Inv}(G, \mathbf{Z}/2\mathbf{Z})$ for G elementary abelian of type $(2,2,\ldots,2)$.

16.1. Cohomology mod 2. We write $H^i(k)$ for $H^i(k, \mathbf{Z}/2\mathbf{Z})$ and $H(k)$ for the direct sum of the $H^i(k)$. Thus $H^0(k) = \mathbf{Z}/2\mathbf{Z}$ and $H^1(k) = k^*/k^{*2}$. For $a \in k^*$, write (a) for the corresponding class in $H^1(k)$, so that $(ab) = (a) + (b)$. The cup product, which we denote by $x \cdot y$, gives $H(k)$ a ring structure. Recall that the cup product $(a)\cdot(b)$ corresponds (via the usual identification of $H^2(k)$ with a subgroup of $\mathrm{Br}(k)$) to the quaternion algebra (a, b); in particular $(a)\cdot(b) = 0$ if and only if the quadratic form $\langle 1, -a, -b\rangle$ represents 0, i.e., b is a norm from the extension $k(\sqrt{a})/k$ or equivalently from the quadratic étale algebra $k[x]/(x^2 - a)$.

16.2. Invariants of $G = \mathbf{Z}/2\mathbf{Z}$. An arbitrary element of $H^1(k, G)$ is given by $\alpha \in k^*/k^{*2}$. Write $\underline{\mathrm{id}}$ for the invariant (i.e., element of $\mathrm{Inv}_{k_0}(\mathbf{Z}/2\mathbf{Z}, \mathbf{Z}/2\mathbf{Z})$) which sends α to $(\alpha) \in H^1(k)$.

Proposition. *$\mathrm{Inv}_{k_0}(\mathbf{Z}/2\mathbf{Z}, \mathbf{Z}/2\mathbf{Z})$ is a free $H(k_0)$-module with basis $\{1, \underline{\mathrm{id}}\}$.*

Proof. Extend scalars to $k_0(t)$ and consider the element $(t) \in H^1(k_0(t))$. It defines a G-torsor T over the projective line minus $\{0, \infty\}$, which from the point of view of quadratic forms corresponds to $\langle t\rangle$. Let $a \in \mathrm{Inv}_{k_0}(\mathbf{Z}/2\mathbf{Z}, \mathbf{Z}/2\mathbf{Z})$ be an invariant; we have $a(T) \in H(k_0(t))$. By 11.7, $a(T)$ is unramified outside $\{0, \infty\}$, hence by 9.4 can be written uniquely as

$$a(T) = \lambda_0 + (t)\cdot\lambda_1 \qquad \text{in } H(k_0(t)) \tag{16.3}$$

with $\lambda_i \in H(k_0)$.

Now define a' to be the invariant $\lambda_0 + \lambda_1 \cdot \underline{\mathrm{id}}$. The invariants a and a' agree on the torsor T, which is obviously versal. Thus the two invariants are equal by 12.3, and 1 and $\underline{\mathrm{id}}$ span $\mathrm{Inv}(\mathbf{Z}/2\mathbf{Z}, \mathbf{Z}/2\mathbf{Z})$.

Note that the element λ_0 occurring in (16.3) is $a(\langle 1\rangle)$, the value of a for the trivial G-torsor over k_0. The other coefficient, λ_1, is the residue at $t = 0$ (or at $t = \infty$) of $a(\langle t\rangle) \in H(k_0(t))$.

Consequently, if a is 0, then $\lambda_0 = \lambda_1 = 0$. That is, the invariants 1 and $\underline{\mathrm{id}}$ are linearly independent over $H(k_0)$. □

The decomposition of an invariant a as

$$a = a_0 + a_1 \cdot \underline{\mathrm{id}}$$

with a_0, $a_1 \in H(k_0)$, is the canonical decomposition of a as the sum (constant) + (normalized), cf. 4.5.

16.4. INVARIANTS OF $G = \mathbf{Z}/2\mathbf{Z} \times \cdots \times \mathbf{Z}/2\mathbf{Z}$. An arbitrary element of $H^1(k,G)$ is given by an n-tuple $(\alpha_1,\dots,\alpha_n) \in k^*/k^{*2} \times \cdots \times k^*/k^{*2}$. For I a subset of $[1,n]$, we write $(\alpha)_I$ for the cup product $\prod_{i\in I}(\alpha_i) \in H(k)$. We have $(\alpha)_\emptyset = 1 \in H^0(k)$.

Write a_I for the invariant $(\alpha_1,\dots,\alpha_n) \mapsto (\alpha)_I$.

THEOREM. *Let $G = \mathbf{Z}/2\mathbf{Z} \times \cdots \times \mathbf{Z}/2\mathbf{Z}$ (n times). Then $\mathrm{Inv}_{k_0}(G,\mathbf{Z}/2\mathbf{Z})$ is a free $H(k_0)$-module with basis $(a_I)_{I\subset[1,n]}$.*

PROOF. By induction on n; the $n=1$ case is Proposition 16.2. Suppose that $n \geq 2$ and write G' for the product of $n-1$ copies of $\mathbf{Z}/2\mathbf{Z}$ so that $G = \mathbf{Z}/2\mathbf{Z} \times G'$, and one writes $(\alpha_1,\dots,\alpha_n)$ as (α_1,α') where α' is an element of $H^1(k,G')$.

Fix such an element $\alpha' \in H^1(k,G')$. Then any $a \in \mathrm{Inv}_{k_0}(G,\mathbf{Z}/2\mathbf{Z})$ defines an invariant $a_{\alpha'} \in \mathrm{Inv}_k(\mathbf{Z}/2\mathbf{Z},\mathbf{Z}/2\mathbf{Z})$ by the formula $a_{\alpha'}(\alpha_1) = a(\alpha_1,\alpha')$. By the $n=1$ case, $a_{\alpha'}(\alpha_1) = \lambda_{0,\alpha'} + (\alpha_1)\lambda_{1,\alpha'}$ with $\lambda_{j,\alpha'} \in H(k)$. When α' varies, the maps $\alpha' \mapsto \lambda_{j,\alpha'}$ are also invariants, i.e., they lie in $\mathrm{Inv}_{k_0}(G',\mathbf{Z}/2\mathbf{Z})$.

By induction, $\lambda_{j,\alpha'} = \sum_{I'\subset[2,n]}(\alpha')_{I'}\lambda_{j,I'}$ for some $\lambda_{j,I'} \in H(k_0)$. Consequently,

$$a(\alpha_1,\alpha') = \Big(\sum(\alpha')_{I'}\cdot\lambda_{0,I'}\Big) + \Big(\sum(\alpha')_{I'}\cdot(\alpha_1)\cdot\lambda_{1,I'}\Big),$$

where the sums run over all subsets I' of $[2,n]$. Define λ_I as $\lambda_{0,I}$ if $1 \notin I$ and as $\lambda_{1,I-\{1\}}$ if $1 \in I$. The above formula can then be rewritten as

$$a(\alpha) = \sum_{I\subset[1,n]} (\alpha)_I\cdot\lambda_I,$$

which shows that the a_I's span $\mathrm{Inv}(G,\mathbf{Z}/2\mathbf{Z})$.

Similarly to what was done in the proof of 16.2, one may recover the λ_I's from a. For $I = \{i_1,i_2,\dots,i_q\}$ a subset of $[1,n]$, define k_I to be the extension of k_0 generated by indeterminates t_i indexed by I. Define an element α_I of $H^1(k_I,G)$ by taking t_i for the i-th coordinate if i belongs to I, and 1 otherwise. Call b_I the value of a on α_I, so b_I is in $H(k_I)$. Take the iterated residue map

$$r_I\colon H(k_I) \to H(k_{I-\{i_1\}}) \to \cdots \to H(k_0),$$

where all residues are taken at 0 (the order in which the residues are taken is irrelevant). Then the λ_I associated with the invariant a is:

$$\lambda_I = r_I(b_I).$$

Thus the a_I's are linearly independent. □

EXERCISE 16.5 (invariants of a product of groups). In this exercise, we write $\mathrm{Inv}_{k_0}(G)$ for $\mathrm{Inv}_{k_0}(G,\mathbf{Z}/p\mathbf{Z})$, where p is a fixed prime.

Let A, A' be algebraic groups over k_0. For every extension k of k_0, we identify $H^1(k,A\times A')$ with the product $H^1(k,A)\times H^1(k,A')$. If $a \in \mathrm{Inv}_{k_0}(A)$ and $a' \in \mathrm{Inv}_{k_0}(A')$, the cup product $a\cdot a'$ is the element of $\mathrm{Inv}_{k_0}(A\times A')$ defined by

$$(a\cdot a')(T,T') = a(T)\cdot a'(T') \quad \text{for every } (T,T') \in H^1(k,A\times A').$$

This defines an $H(k_0)$-linear map

$$c_{A,A'}\colon \mathrm{Inv}_{k_0}(A)\otimes\mathrm{Inv}_{k_0}(A') \to \mathrm{Inv}_{k_0}(A\times A'),$$

where the tensor product is taken over $H(k_0) = \oplus_i H^i(k_0,\mathbf{Z}/p\mathbf{Z})$.

Make the following assumption on A:

(*) There exists a family (a_j) of elements of $\mathrm{Inv}_{k_0}(A)$ such that, for every k/k_0, the images of the a_j in $\mathrm{Inv}_k(A)$ make up an $H(k)$-basis of $\mathrm{Inv}_k(A)$.

Prove that $c_{A,A'}$ is then an isomorphism.

[Hint: Surjectivity. Let a be an element of $\mathrm{Inv}_{k_0}(A \times A')$. If T' is an element of $H^1(k, A')$, the map $T \mapsto a(T, T')$ is an element of $\mathrm{Inv}_k(A)$, hence can be written in a unique way as $\sum a_j \otimes b_j(T')$, with $b_j(T') \in H(k)$. The map $T' \mapsto b_j(T')$ is an element b_j of $\mathrm{Inv}_{k_0}(A')$, and it is clear that $c_{A,A'}(\sum a_j \otimes b_j)$ is equal to a. Injectivity is proved similarly.]

Show that, if A' also has property (*), so does $A \times A'$. Note that the case $p = 2$ gives an alternate proof of Th. 16.4, by induction on n. Note also that assumption (*) is satisfied (when $p = 2$) by several of the groups considered in later sections, such as $\mathbf{O}_n$, $\mathbf{SO}_n$ (n odd), G_2, F_4, E_6 (split), and the symmetric group S_n.

17. Cohomological invariants of Quad_n

In this section we determine $\mathrm{Inv}(\mathrm{Quad}_n, \mathbf{Z}/2\mathbf{Z})$.

17.1. Stiefel-Whitney classes. If q is a quadratic form of rank n over k, we may write it as $q = \langle \alpha_1, \alpha_2, \ldots, \alpha_n \rangle$ for $\alpha_i \in k^*$. Let $w_i(q)$ be the i-th elementary symmetric polynomial in the (α_j)'s computed in the commutative ring $H(k)$:

$$w_0 = 1,$$
$$w_1 = \sum_i (\alpha_i) = (\alpha_1 \alpha_2 \cdots \alpha_n) = (d(q)),$$
$$w_2 = \sum_{i<j} (\alpha_i) \cdot (\alpha_j),$$
$$\ldots,$$
$$w_n = (\alpha_1) \cdot (\alpha_2) \cdot \cdots \cdot (\alpha_n),$$
$$\text{and } w_i = 0 \text{ if } i < 0 \text{ or } i > n.$$

It is known (cf. [Delz 62] or [Mi 70]) that these are indeed invariants of q, i.e., they do not depend on the particular way of writing q as $\langle \alpha_1, \ldots, \alpha_n \rangle$. They provide $n + 1$ elements in $\mathrm{Inv}(\mathrm{Quad}_n, \mathbf{Z}/2\mathbf{Z})$.

17.2. The total Stiefel-Whitney class. We write $w(q)$ for the *total Stiefel-Whitney class*

$$w(q) = \sum_{i=0}^{n} w_i(q) \quad \text{in } H(k).$$

It is characterized by the properties

(1) $w(\langle \alpha \rangle) = 1 + (\alpha)$
(2) $w(q \oplus q') = w(q) \cdot w(q')$.

17.3. Invariants of Quad_n.

Theorem. *The group* $\mathrm{Inv}_{k_0}(\mathrm{Quad}_n, \mathbf{Z}/2\mathbf{Z})$ *is a free* $H(k_0)$*-module with basis* $\{w_0, w_1, \ldots, w_n\}$.

Proof. Let $G = \mathbf{Z}/2\mathbf{Z} \times \cdots \times \mathbf{Z}/2\mathbf{Z}$ as in Section 16. An arbitrary element of $H^1(k, G)$ is given by an n-tuple

$$(\alpha_1, \ldots, \alpha_n) \in k^*/k^{*2} \times \cdots \times k^*/k^{*2}.$$

The map $(\alpha_1, \ldots, \alpha_n) \mapsto \langle \alpha_1, \ldots, \alpha_n \rangle$ defines a morphism from the functor $H^1(*, G)$ to the functor Quad_n.

Thus any invariant $a \in \mathrm{Inv}(\mathrm{Quad}_n, \mathbf{Z}/2\mathbf{Z})$ gives an invariant a_G in $\mathrm{Inv}(G, \mathbf{Z}/2\mathbf{Z})$; one may view a_G as the restriction of a relative to the diagonal embedding $G \to \mathbf{O}_n$, cf. §13. By Theorem 16.4, a_G is of the form

$$a_G(\alpha_1, \ldots, \alpha_n) = \sum_{I \subset [1,n]} \lambda_I \cdot (\alpha)_I \quad \text{for } \lambda_I \in H(k),$$

where $(\alpha)_I$ is, as above, the cup product of the (α_i) for $i \in I$. But since a is an invariant of Quad_n, it is unchanged when we permute the α_i's. Consequently, the value of λ_I depends only on the cardinality $|I|$ of I. Since $\sum_{|I|=i} (\alpha)_I = w_i(q)$, a is of the desired form. Thus the Stiefel-Whitney classes span $\mathrm{Inv}(\mathrm{Quad}_n, \mathbf{Z}/2\mathbf{Z})$.

If a is 0, then by the linear independence of the $(\alpha)_I$'s in $\mathrm{Inv}(G, \mathbf{Z}/2\mathbf{Z})$ (16.4) all of the λ_I's must be 0. Consequently the Stiefel-Whitney classes are linearly independent. (Alternately, one may also extract the λ_I's from a as in the proof of 16.4.) □

The proof shows that the natural restriction map

$$\mathrm{Inv}(\mathbf{O}_n, \mathbf{Z}/2\mathbf{Z}) \to \mathrm{Inv}(G, \mathbf{Z}/2\mathbf{Z})$$

is injective and that its image is the subgroup of $\mathrm{Inv}(G, \mathbf{Z}/2\mathbf{Z})$ made up of the elements fixed under the natural action of S_n on G (i.e., fixed under the action of the normalizer of G in $\mathbf{O}_n$, cf. 13.2).

REMARK 17.4. A consequence of Theorem 17.3 is:

(1) The cup product of two Stiefel-Whitney classes can be written as a linear combination of the Stiefel-Whitney classes, since such a product is an invariant. An explicit formula can be found in [Mi 70, p. 331, line 2]: To compute $w_r \cdot w_s$, write r in dyadic form as $r = \sum_{i \in R} 2^i$ for $R \subset \{0, 1, 2, \ldots\}$ and similarly write $s = \sum_{i \in S} 2^i$. Then

$$w_r(q) \cdot w_s(q) = \varepsilon^m \cdot w_{r+s-m}(q), \quad \text{where } m = \sum_{i \in R \cap S} 2^i$$

and $\varepsilon = (-1) \in H^1(k)$, the powers ε^m of ε being relative to the cup product. Equation (19.3) is the special case $r = 1$, where $m = 0$ if s is even and $m = 1$ if s is odd.

This shows in particular (cf. [Mi 70]) that every w_r can be written as a monomial in $w_1, w_2, w_4, w_8, \ldots$ More precisely, if $r = \sum_{i \in R} 2^i$ as above, then $w_r(q) = \prod_{i \in R} w_{2^i}(q)$.

(2) There are exterior powers $\lambda^p q$ associated with every quadratic form q. (See 27.1 for a definition and an explicit formula.) Their Stiefel-Whitney classes can be written (by a universal formula) as a linear combination of the Stiefel-Whitney classes of q. An example of this is

$$w_1(\lambda^2 q) = \begin{cases} 0 & \text{if } \mathrm{rank}(q) \text{ is odd} \\ w_1(q) & \text{if } \mathrm{rank}(q) \text{ is even.} \end{cases}$$

EXERCISE 17.5. (1) Let x, $y \in k^*$ with $x + y \neq 0$. Show that

$$(x)\cdot(y) = (x+y)\cdot(-xy) \quad \text{in } H^2(k)$$

by proving that the quadratic forms $\langle x, y\rangle$ and $\langle x+y, xy(x+y)\rangle$ are isomorphic and computing their w_2 invariant.

(2) Let x, y, $z \in k^*$. Put

$$c_1 = x + y + z, \quad c_2 = xy + yz + zx, \quad c_3 = xyz.$$

Prove the following identities in $H^2(k)$ and $H^3(k)$:
(2a) $(x)\cdot(y) + (y)\cdot(z) + (z)\cdot(x) = (c_1)\cdot(-c_3) + (c_1c_2 - 9c_3)\cdot(-c_1c_3)$ and
(2b) $(x)\cdot(y)\cdot(z) = (c_1)\cdot(c_3)\cdot(c_1c_2 - 9c_3)$
under the assumption that $c_1 \neq 0$ and $c_1c_2 \neq 9c_3$.

[Hint: Let $E = k \times k \times k$, and let q be the diagonal quadratic form on E defined by $\langle x, y, z\rangle$. The left side of (2a) is $w_2(q)$. Define e_1, $e_2 \in E$ by $e_1 = (1,1,1)$ and $e_2 = (y-z, z-x, x-y)$. The vectors e_1 and e_2 are orthogonal and anisotropic. Choose e_3 orthogonal to e_1, e_2 and nonzero. If one computes $w_2(q)$ using the basis $\{e_1, e_2, e_3\}$, one finds the right side of (2a). Identity (2b) is deduced from (2a) by multiplication by $(c_3) = (x) + (y) + (z)$.]

(3) Let $k = k_0(x_1, \ldots, x_n)$, where the x_i are independent indeterminates, and let c_1, …, c_n be the elementary symmetric polynomials in the x_i. Show that there exists a quadratic form over $k_0(c_1, \ldots, c_n)$ which becomes isomorphic to $\langle x_1, \ldots, x_n\rangle$ over k. Deduce that any symmetric polynomial in (x_1), …, (x_n) can be computed by universal polynomial formulas from the symmetric polynomials in the x_i's, just as above for $n \leq 3$.

18. Cohomological invariants of Pfister$_n$ and Oct

We now determine $\mathrm{Inv}(\mathrm{Pfister}_n, \mathbf{Z}/2\mathbf{Z})$. An n-Pfister form may be written as

$$\phi = \langle 1, -\alpha_1\rangle \otimes \cdots \otimes \langle 1, -\alpha_n\rangle.$$

Then

$$e_n(\phi) = (\alpha_1)\cdot(\alpha_2)\cdots(\alpha_n) \in H^n(k)$$

is an invariant of ϕ by [EL 72, 3.2].

THEOREM 18.1. *The group $\mathrm{Inv}_{k_0}(\mathrm{Pfister}_n, \mathbf{Z}/2\mathbf{Z})$ of cohomological invariants of n-Pfister forms is a free $H(k_0)$-module with basis $\{1, e_n\}$.*

PROOF. For simplicity of notation, we give the proof only for the case $n = 2$. A similar argument works for any n.

Let $\phi = \langle 1, -\alpha\rangle \otimes \langle 1, -\beta\rangle \in \mathrm{Pfister}_2(k)$. Any $a \in \mathrm{Inv}(\mathrm{Pfister}_2, H)$ gives by restriction an invariant of the pair (α, β), where α and β are viewed as one-dimensional quadratic forms, or as elements of k^*/k^{*2}, or as $\mathbf{Z}/2\mathbf{Z}$-torsors. By Theorem 16.4 a is of the form

$$a(\phi) = \lambda_0 + (\alpha)\cdot\lambda_1 + (\beta)\cdot\lambda_2 + (\alpha)\cdot(\beta)\cdot\lambda_{12} \tag{18.2}$$

where the λ's belong to $H(k_0)$. Since α and β may be swapped without changing ϕ, we have $\lambda_1 = \lambda_2$. Also, $\phi \cong \langle 1, -\alpha\rangle \otimes \langle 1, \alpha\beta\rangle$, so

$$a(\phi) = \lambda_0 + (\alpha)\cdot\lambda_1 + (-\alpha\beta)\cdot\lambda_1 + (\alpha)\cdot(-\alpha\beta)\cdot\lambda_{12}. \tag{18.3}$$

By adding (18.2) and (18.3) we get $(-\alpha)\cdot\lambda_1 + (\alpha)\cdot(-\alpha)\cdot\lambda_{12} = 0$. Since $(\alpha)\cdot(-\alpha) = 0$, we have $(-\alpha)\cdot\lambda_1 = 0$. This is true for every extension k of k_0 and for every $\alpha \in k^*$; that is, the invariant

$$b = (-1)\cdot\lambda_1 + \underline{\mathrm{id}}\cdot\lambda_1 \qquad \text{in } \mathrm{Inv}_{k_0}(\mathbf{Z}/2\mathbf{Z}, \mathbf{Z}/2\mathbf{Z})$$

is 0. Since 1 and $\underline{\mathrm{id}}$ are linearly independent in $\mathrm{Inv}_{k_0}(\mathbf{Z}/2\mathbf{Z}, \mathbf{Z}/2\mathbf{Z})$ by 16.2, we have $\lambda_1 = 0$. This proves the theorem for $n = 2$. □

18.4. OCTONIONS. An octonion algebra O has a norm polynomial q_O, which is a 3-Pfister (cf. 2.8), and the invariant of this 3-Pfister is an element $e(O)$ of $H^3(k)$. What 18.1 shows is that $\mathrm{Inv}_{k_0}(\mathrm{Oct}, \mathbf{Z}/2\mathbf{Z})$ *is a free* $H(k_0)$*-module with basis* $\{1, e\}$. In particular, every normalized invariant is a multiple of the e-invariant.

If O has basis $\{1, e_1, e_2, \ldots, e_7\}$ as usual, with

$$\begin{aligned} e_1e_1 &= \alpha_1, & e_2e_2 &= \alpha_2, & e_3e_3 &= \alpha_3, \\ e_4 = e_1e_2 &= -e_2e_1, & e_5 = e_2e_3 &= -e_3e_2, \\ e_6 = e_3e_4 &= -e_4e_3, & e_7 = e_4e_5 &= -e_5e_4, \end{aligned}$$

then the form q_O is

$$q_O \cong \langle 1, -\alpha_1\rangle \otimes \langle 1, -\alpha_2\rangle \otimes \langle 1, -\alpha_3\rangle,$$

and the e-invariant of O is the cup product $(\alpha_1)\cdot(\alpha_2)\cdot(\alpha_3) \in H^3(k)$.

19. Cohomological invariants of $\mathrm{Quad}_{n,\delta}$ (n odd)

Fix $\delta \in k_0^*/k_0^{*2}$, and recall that $\mathrm{Quad}_{n,\delta}(k)$ is the set of isomorphism classes of nondegenerate quadratic forms of rank n and discriminant δ over k as in 2.5.

THEOREM 19.1. *For n odd,* $\mathrm{Inv}_{k_0}(\mathrm{Quad}_{n,\delta}, \mathbf{Z}/2\mathbf{Z})$ *is a free* $H(k_0)$*-module with basis* $\{w_0, w_2, w_4, \ldots, w_{n-1}\}$.

PROOF. We use the following two formulas, which are proved by direct computation: For q a quadratic form,

$$w_i(q \oplus \langle\alpha\rangle) = (\alpha)\cdot w_{i-1}(q) + w_i(q) \tag{19.2}$$

and

$$w_1(q)\cdot w_{i-1}(q) = \begin{cases} w_i(q) & \text{if } i \text{ is odd} \\ (-1)\cdot w_{i-1}(q) & \text{if } i \text{ is even.} \end{cases} \tag{19.3}$$

If q_1 is any quadratic form of rank $n-1$, we can associate with it a rank n form q with discriminant δ by the recipe:

$$q = q_1 \oplus \langle d(q_1)\delta\rangle. \tag{19.4}$$

Hence every invariant a of $\mathrm{Quad}_{n,\delta}$ defines an invariant a_1 of Quad_{n-1} by the formula $a_1(q_1) = a(q)$. Note that $a_1 = \mathrm{Res}(a)$, where the restriction map (§13) is relative to the natural embedding $\mathbf{O}_{n-1} \to \mathbf{SO}(Q_\delta)$, where $Q_\delta = \langle 1, \ldots, 1, \delta\rangle$.

By Theorem 17.3, there are well-defined elements λ_0, ..., λ_{n-1} of $H(k_0)$ such that

$$a_1(q_1) = \sum_{i=0}^{n-1} \lambda_i\cdot w_i(q_1). \tag{19.5}$$

For q in $\mathrm{Quad}_{n,\delta}$, we may write

$$q = \langle \alpha_1, \alpha_2, \ldots, \alpha_{n-1}, \delta\alpha_1\alpha_2\cdots\alpha_{n-1}\rangle.$$

Set $q_0 = \langle\alpha_1, \ldots, \alpha_{n-2}\rangle$ and $q_1 = q_0 \oplus \langle\alpha_{n-1}\rangle$, so that (19.4) holds. We have

$$a(q) = a_1(q_1) = a_1(q_0 \oplus \langle\alpha_{n-1}\rangle) = \sum_{i=0}^{n-1} \lambda_i\cdot w_i(q_0 \oplus \langle\alpha_{n-1}\rangle). \tag{19.6}$$

The value of $a(q)$ must remain unchanged if we write q differently, in particular if we swap the last two entries (that is, if we interchange the terms α_{n-1} and $\delta\alpha_1 \dots \alpha_{n-1}$). Thus

$$(19.7) \qquad \begin{aligned} a(q) &= a_1(q_0 \oplus \langle \delta\alpha_1 \cdots \alpha_{n-1}\rangle) \\ &= \textstyle\sum_{i=0}^{n-1} \lambda_i \cdot w_i(q_0 \oplus \langle \delta\alpha_1 \cdots \alpha_{n-1}\rangle). \end{aligned}$$

The difference of (19.6) and (19.7) gives by (19.2):

$$\sum_{i=0}^{n-1} \lambda_i \cdot (\delta\alpha_1 \dots \alpha_{n-2}) \cdot w_{i-1}(q_0) = 0.$$

We have $(\delta\alpha_1 \dots \alpha_{n-2}) = (\delta) + w_1(q_0)$, and applying (19.3) to expand the product $w_1(q_0)w_{i-1}(q_0)$, we obtain

$$(19.8) \qquad \sum_{\substack{i=0 \\ i \text{ even}}}^{n-1} \lambda_i \cdot (-\delta) \cdot w_{i-1}(q_0) + \sum_{\substack{i=0 \\ i \text{ odd}}}^{n-1} \lambda_i \cdot [(\delta) \cdot w_{i-1}(q_0) + w_i(q_0)] = 0.$$

Since $\{w_0, w_1, \dots, w_{n-2}\}$ is a basis for $\mathrm{Inv}(\mathrm{Quad}_{n-2}, \mathbf{Z}/2\mathbf{Z})$ by 17.3, the coefficients of these w_i must all be 0. Examining the coefficients of the Stiefel-Whitney classes of even degree, we have

$$(19.9) \qquad \lambda_i \cdot (\delta) = 0 \qquad \text{for } i \text{ odd, } 1 \le i < n.$$

Collecting the coefficients of the classes of odd degree, we obtain

$$(19.10) \qquad \lambda_i = \lambda_{i+1} \cdot (-\delta) \qquad \text{for } i \text{ odd, } 1 \le i < n.$$

For i even (including the case where $i = 0$, where we set $\lambda_{-1} = 0$) we have:

$$(19.11) \qquad \begin{aligned} &\lambda_{i-1} \cdot w_{i-1}(q_1) + \lambda_i \cdot w_i(q_1) = \\ &\quad = \lambda_i \cdot [(-1) \cdot w_{i-1}(q_1) + (\delta) \cdot w_{i-1}(q_1) + w_i(q_1)] && \text{by (19.10)} \\ &\quad = \lambda_i \cdot [w_1(q_1) w_{i-1}(q_1) + (\delta) \cdot w_{i-1}(q_1) + w_i(q_1)] && \text{by (19.3)} \\ &\quad = \lambda_i \cdot [(\delta\alpha_1 \cdots \alpha_{n-1}) \cdot w_{i-1}(q_1) + w_i(q_1)] && \text{by (19.4)} \\ &\quad = \lambda_i \cdot w_i(q) && \text{by (19.2)} \end{aligned}$$

Combining this with (19.5) and using the fact that n is odd, we obtain:

$$(19.12) \qquad a(q) = \sum_{\substack{i=0 \\ i \text{ even}}}^{n-1} \lambda_i \cdot w_i(q).$$

This proves that the even Stiefel-Whitney classes span $\mathrm{Inv}(\mathrm{Quad}_{n,\delta}, \mathbf{Z}/2\mathbf{Z})$.

Suppose now that the invariant a given by (19.12) is 0. Then so is a_1. Since the Stiefel-Whitney classes are linearly independent in $\mathrm{Inv}(\mathrm{Quad}_{n-1}, \mathbf{Z}/2\mathbf{Z})$ by 17.3, the λ_i must be 0 for all i, hence in particular for i even. Thus the even Stiefel-Whitney classes $\{w_0, w_2, \dots, w_{n-1}\}$ are linearly independent in $\mathrm{Inv}(\mathrm{Quad}_{n,\delta}, \mathbf{Z}/2\mathbf{Z})$. □

EXERCISE 19.13. For n odd, define an isomorphism of functors

$$\mathrm{Quad}_{n,\delta} \times \mathrm{Quad}_1 \to \mathrm{Quad}_n$$

by $(q, q_1) \mapsto q \otimes q_1$, and use Exercise 16.5 to recover Theorem 19.1.

20. Cohomological invariants of $\mathrm{Quad}_{n,\delta}$ (n even)

PROPOSITION 20.1. *Assume that n is even, $n > 2$; let $\lambda \in H(k_0)$ be such that $(\delta)\cdot\lambda = 0$ in $H(k_0)$. Let q be an element of $\mathrm{Quad}_{n,\delta}(k)$. Write q as $\langle\alpha_1, \dots, \alpha_n\rangle$. Then the element $(\alpha_1)\cdot(\alpha_2)\cdots(\alpha_{n-1})\cdot\lambda$ of $H(k)$ depends only on λ and the isomorphism class of q. If we denote it by $b_\lambda(q)$, the map $q \mapsto b_\lambda(q)$ is an invariant*

$$\mathrm{Quad}_{n,\delta} \to H.$$

[If $\delta = 1$, the condition $\lambda\cdot(\delta) = 0$ is empty, so b_λ is defined for every λ and is equal to the cup product $b_1\cdot\lambda$.]

If $q = \langle\alpha_1, \dots, \alpha_n\rangle$ has discriminant δ, we write $\underline{\alpha}$ for the n-tuple $(\alpha_1, \dots, \alpha_n)$ and $c(q, \underline{\alpha})$ for the image of $(\alpha_1)\cdots(\alpha_{n-1})$ in the quotient $H^{n-1}(k)/(\delta)\cdot H^{n-2}(k)$. Since multiplication by λ defines a homomorphism of $H^{n-1}(k)/(\delta)\cdot H^{n-2}(k)$ into $H(k)$, the following lemma implies Proposition 20.1:

LEMMA 20.2. *$c(q, \underline{\alpha})$ depends only on q* (but not on $\underline{\alpha}$).

PROOF. The proof is in two steps:

i) $c(q, \underline{\alpha})$ does not change when the α_i are permuted.

It is enough to show this for a transposition $(i\,j)$. If both i and j are $< n$, the invariance of $c(q, \underline{\alpha})$ is clear. If, say, $i = n$ and $j < n$, we have to show that $x\cdot(\alpha_j) = x\cdot(\alpha_n)$ in $H^{n-1}(k)/(\delta)\cdot H^{n-2}(k)$, where x is the cup product of the (α_r) for $r \neq j, n$. By assumption, we have $\alpha_1 \cdots \alpha_n = \delta$ in k^*/k^{*2}, hence (using the fact that n is even):

$$(\alpha_n) = (\delta) + (\alpha_j) + \sum_{r \neq j,n} (-\alpha_r).$$

Since $x\cdot(-\alpha_r) = 0$ for all $r \neq j, n$, this gives by multiplying with x:

$$x\cdot(\alpha_j) \equiv x\cdot(\alpha_n) \mod (\delta)\cdot H^{n-2}(k),$$

as desired.

ii) We have $c(q, \underline{\alpha}) = c(q, \underline{\beta})$ if $q \cong \langle\beta_1, \dots, \beta_n\rangle$.

By Witt's chain equivalence theorem [Lam 73, I.5.2] it is enough to prove this when $\beta_i = \alpha_i$ for all indices i except two of them, say j and j'. By part i) (and the hypothesis $n \geq 4$) we may assume that j and j' are both $< n$. But in that case we have $(\beta_j)\cdot(\beta_{j'}) = (\alpha_j)\cdot(\alpha_{j'})$ since the quadratic forms $\langle\beta_j, \beta_{j'}\rangle$ and $\langle\alpha_j, \alpha_{j'}\rangle$ are isomorphic. Hence $c(q, \underline{\beta}) = c(q, \underline{\alpha})$. □

Note that the definition of b_λ could have been given differently: choose $\alpha \neq 0$ represented by q, and write q as $q'_\alpha \oplus \langle\alpha\rangle$. Then b_λ is the image of $w_{n-1}(q'_\alpha)\cdot\lambda$ in $H(k)$. Lemma 20.2 says that b_λ does not depend on the choice of α.

EXAMPLE 20.3. Assume $n = 4$, $\delta = 1$. Choose α represented by q. Then $q_2 = \langle\alpha\rangle q$ is a 2-Pfister form, and $q_3 = \langle 1, -\alpha\rangle q_2$ is a 3-Pfister form. The form q_2 is independent of the choice of α, as follows from the computation of w_2 or from Lemma 22.2 below. We have (in the Witt-Grothendieck ring of k, cf. 27.1):

$$q = \langle -1\rangle(q_3 - q_2),$$

an equation which determines q_2 and q_3 completely (Lemma 22.2). The b_1 invariant of q is then:

$$b_1(q) = e_3(q_3).$$

It follows from a theorem of Merkurjev (see e.g. [Ar 84, Prop. 2]) that q is uniquely determined by its invariants $w_2(q) \in H^2(k)$ and $b_1(q) \in H^3(k)$. In particular, q is a 2-Pfister form if and only if $b_1(q) = 0$.

(For a more direct description of b_1 in the $\delta = 1$ case, see the letter of M. Rost reproduced in Appendix A.)

20.4. Let I_δ denote the ideal in $H(k_0)$ consisting of those λ such that $(\delta)\cdot\lambda = 0$. For n even and ≥ 4, there is a map

$$I_\delta \longrightarrow \mathrm{Inv}_{k_0}(\mathrm{Quad}_{n,\delta}, \mathbf{Z}/2\mathbf{Z}) \tag{20.5}$$

of $H(k_0)$-modules given by $\lambda \mapsto b_\lambda$, cf. Prop. 20.1.

Let $\lambda \in I_\delta$ lie in the kernel of (20.5). For $q_1 \in \mathrm{Quad}_{n-1}$, we have

$$0 = b_\lambda(q + \langle d(q_1)\delta\rangle) = \lambda\cdot w_{n-1}(q_1).$$

By 17.3, this implies $\lambda = 0$. That is, the map in (20.5) is injective.

THEOREM 20.6. *For n even and ≥ 4, $\mathrm{Inv}_{k_0}(\mathrm{Quad}_{n,\delta}, \mathbf{Z}/2\mathbf{Z})$ is a direct sum of the free $H(k_0)$-module with basis $\{w_0, w_2, \ldots, w_{n-2}\}$ and the image of the ideal I_δ.*

PROOF. The proof of this theorem is the same as the proof of 19.1 up to (19.9) and (19.10). For $i = n-1$ (which is odd), these formulas are not immediately implied by (19.8) since $w_{n-1}(q_0) = 0$. Instead, examining the coefficient of $w_{n-1}(q_0)$ in (19.8), we have

$$\lambda_{n-1}\cdot(\delta) = 0. \tag{20.7}$$

Equation (19.11) holds with no change, and instead of (19.12), we obtain

$$a(q) = \sum_{\substack{i=0\\ i \text{ even}}}^{n-2} \lambda_i\cdot w_i(q) + \lambda_{n-1}\cdot w_{n-1}(q_1). \tag{20.8}$$

The second term in the sum may be written as $b_{\lambda_{n-1}}(q)$ by (20.7), which proves that the specified elements span $\mathrm{Inv}(\mathrm{Quad}_{n,\delta}, \mathbf{Z}/2\mathbf{Z})$.

Suppose now that the invariant a given by (20.8) is 0. Then the induced invariant on Quad_{n-1} is 0 and the linear independence (17.3) of the Stiefel-Whitney classes in $\mathrm{Inv}(\mathrm{Quad}_{n-1}, \mathbf{Z}/2\mathbf{Z})$ implies that all of the coefficients λ_i are 0. Thus, the set $\{1, w_2, w_4, \ldots, w_{n-2}\}$ is linearly independent, and its span has intersection $\{0\}$ with the image of I_δ. □

EXERCISE 20.9 (the case $n = 2$). Let J_δ be the ideal of $H(k_0)$ generated by $(-\delta)$. It is contained in the ideal I_δ of 20.4. Write q in $\mathrm{Quad}_{2,\delta}$ as $\langle\alpha_1, \alpha_2\rangle$. For $\lambda \in J_\delta$:

(1) Show that $(\alpha_1)\cdot\lambda = (\alpha_2)\cdot\lambda$.
(2) Show that $(\alpha_1)\cdot\lambda$ only depends on q.

This gives an invariant c_λ for each $\lambda \in J_\delta$.

(3) Prove that $\mathrm{Inv}_{k_0}(\mathrm{Quad}_{2,\delta}, \mathbf{Z}/2\mathbf{Z})$ is the direct sum of the constant invariants and the ideal J_δ.

(Note that I_δ is the image of Cor : $H(k(\sqrt{\delta})) \to H(k)$ and that J_δ is the kernel of $H(k) \to H(k(\sqrt{-\delta}))$, cf. [Ar 75, Cor. 4.6])

21. Cohomological invariants of hermitian forms

Let $k_1 = k_0(\sqrt{\delta})$ be a quadratic extension of k_0. We will determine the group $\mathrm{Inv}(\mathrm{Herm}_n, \mathbf{Z}/2\mathbf{Z})$.

21.1. RELATIONS WITH Quad_n AND Quad_{2n}. There is a natural map

$$\mathrm{Quad}_n \to \mathrm{Herm}_n,$$

namely $\langle \alpha_1, \ldots, \alpha_n \rangle \mapsto \langle \alpha_1, \ldots, \alpha_n \rangle_H$. Since every hermitian form may be diagonalized to $\langle \alpha_1, \ldots, \alpha_n \rangle_H$ with $\alpha_i \in k^*$, this map is *surjective.*

Let V be a free module of rank n over the quadratic algebra $k \otimes_{k_0} k_1$, and let h be a nondegenerate hermitian form on V. Then V is a $2n$-dimensional k-vector space endowed with a quadratic form $v \mapsto h(v,v)$; we denote this quadratic form by Nh. This gives a map $N\colon \mathrm{Herm}_n \to \mathrm{Quad}_{2n}$ which is well-known to be *injective.* Hence we may view $\mathrm{Herm}_n(k)$ as a subset of $\mathrm{Quad}_{2n}(k)$. (When $\sqrt{\delta} \in k$, this subset is $\{\mathrm{hyp}\}$, where hyp is hyperbolic of rank $2n$; in the general case it is the set of $q \in \mathrm{Quad}_{2n}(k)$ which become hyperbolic over $k(\sqrt{\delta})$.)

The map

$$\mathrm{Quad}_n(k) \to \mathrm{Herm}_n(k) \xrightarrow{N} \mathrm{Quad}_{2n}(k) \tag{21.2}$$

is $q \mapsto \langle 1, -\delta \rangle q$; i.e., $N(q_H) = \langle 1, -\delta \rangle \cdot q$.

21.3. STIEFEL-WHITNEY CLASSES. Let $x \in H(k_0)$ be an element which gives 0 in $H(k_1)$, i.e., which belongs to the ideal $(\delta)\cdot H(k_0)$ generated by (δ) [Ar 75, Cor. 4.6]. Let i be an integer ≥ 1. If $h \in \mathrm{Herm}_n(k)$, choose $q \in \mathrm{Quad}_n(k)$ which gives h as in 21.1, i.e., such that $q_H = h$.

LEMMA 21.4. *The cup product $x \cdot w_i(q) \in H(k)$ depends only on x and h* (and not on the choice of q).

PROOF. It is enough to prove the lemma when $x = (\delta)$. When δ is a square in k, we have $x = 0$ and the statement is obvious. We may thus assume that δ is not a square, so that the field $K = k(\sqrt{\delta})$ is a quadratic extension of k. We now consider three cases:

i) The case $i = 1$. If $q = \langle \alpha_1, \ldots, \alpha_n \rangle$, we have $w_1(q) = (\alpha_1 \cdots \alpha_n)$. The hermitian form $h = q_H$ has determinant $d \in k^*/NK^*$, which is the image of $\alpha_1 \cdots \alpha_n$ in k^*/NK^* (see [KMRT 98, p. 114]). The natural embedding $k^*/NK^* \to H^2(k)$ transforms d into $(\delta)\cdot(d) = (\delta)\cdot w_1(q)$. This shows that $(\delta)\cdot w_1(q)$ depends only on h.

ii) The case $n \leq 2$. Since $w_i(q) = 0$ if $i > n$, we may assume that $i \leq n$. By part i), we may also assume that $i > 1$. This only leaves the case $i = 2$, $n = 2$. The form $Nh \in \mathrm{Quad}_4(k)$ is $\langle \alpha_1, -\alpha_1\delta, \alpha_2, -\alpha_2\delta \rangle$. Its discriminant is 1, and its b_1 invariant (in the sense of 20.3) is $(\alpha_1)\cdot(\alpha_2)\cdot(-\alpha_1\delta) = (\alpha_1)\cdot(\alpha_2)\cdot(\delta) = (\delta)\cdot w_2(q)$. This shows that $(\delta)\cdot w_2(q) = b_1(Nh)$ depends only on $h = q_H$.

iii) The case $n > 2$. This follows from the case $n = 2$ by Witt's chain equivalence for hermitian forms (cf. [Wi 37, p. 36] — note that the proof for quadratic forms in [Lam 73, I.5.2] extends to the hermitian case with practically no change). □

Thanks to Lemma 21.4, for every $x \in (\delta)\cdot H(k)$ we may define a cohomological invariant $w_{i,x} \in \mathrm{Inv}(\mathrm{Herm}_n, \mathbf{Z}/2\mathbf{Z})$ by putting

$$w_{i,x}(h) = x \cdot w_i(q)$$

where $q \in \mathrm{Quad}_n(k)$ is such that $q_H = h$.

REMARK 21.5. Note that, in the case $n = 2$, an element of $\mathrm{Herm}_n(k)$ is uniquely determined by its two invariants $w_{1,(\delta)} \in H^2(k)$ and $w_{2,(\delta)} \in H^3(k)$; this follows from 20.3.

THEOREM 21.6. *Every element of* $\mathrm{Inv}_{k_0}(\mathrm{Herm}_n, \mathbf{Z}/2\mathbf{Z})$ *can be written uniquely as*

$$h \mapsto x_0 + \sum_{i=1}^{n} w_{i,x_i}(h)$$

with $x_0 \in H(k_0)$ *and* $x_i \in (\delta)\cdot H(k_0)$ *for* $i \geq 1$.

PROOF. Any invariant of Herm_n gives one of Quad_n by 21.1, which can be written uniquely as

$$q \mapsto x_0 + \sum_{i=1}^{n} x_i \cdot w_i(q), \text{ if } q \mapsto h, \text{ with } x_i \in H(k_0).$$

Moreover, this invariant should give a *constant* invariant over k_1. This shows that the x_i $(i \geq 1)$ belong to $(\delta)\cdot H(k_0) = \ker[H(k_0) \to H(k_1)]$. □

22. Cohomological invariants of Albert algebras

We will now determine $\mathrm{Inv}(\mathrm{Alb}, \mathbf{Z}/2\mathbf{Z})$, cf. 2.9.

EXAMPLE 22.1 (reduced Albert algebras). Let q_2 and q_3 be 2- and 3-Pfister forms over k. Let O be an octonion algebra with norm q_3. Choose α_1 and α_2 in k^* such that q_2 is isomorphic to $\langle 1, -\alpha_1\rangle \otimes \langle 1, -\alpha_2\rangle$ and set

$$g = \begin{pmatrix} -\alpha_1 & & \\ & -\alpha_2 & \\ & & \alpha_1\alpha_2 \end{pmatrix}$$

in $M_3(O)$. There is a k-linear map $*$ defined for $x \in M_3(O)$ by $x^* = g\cdot\overline{x}^t\cdot g^{-1}$, where t is the transpose, and $^-$ is the canonical conjugation on O. The subspace of $M_3(O)$ consisting of those x with $x = x^*$ endowed with the product $(x, y) \mapsto (xy + yx)/2$ is an Albert algebra which we denote by J. It is determined up to isomorphism by q_2 and q_3.

This construction gives a map $\mathrm{Oct} \times \mathrm{Pfister}_2 \to \mathrm{Alb}$ corresponding to the natural inclusion

$$G_2 \times PGL_2 \to F_4.$$

The Albert algebras obtained from this construction are called *reduced*, see [SV 00, p. 125], [KMRT 98, p. 517], or [PR 94].

LEMMA 22.2. *Suppose that* u_i *is an odd-dimensional quadratic form and* f_i, f'_i *are* i*-Pfister forms for some finite set of indices* i. *If* $\sum u_i f_i = \sum u_i f'_i$ *in the Witt ring of* k, *then* $f_i = f'_i$ *for all* i.

PROOF. Write I for the augmentation ideal of the Witt ring $W(k)$, cf. 27.1. Let n be the smallest of the indices. Modulo I^{n+1}, $\sum u_i f_i$ is congruent to $u_n f_n$, which is in turn congruent to f_n, and similarly for $\sum u_i f'_i$. Since f_n and f'_n are n-Pfister forms which are congruent modulo I^{n+1}, they are isomorphic by a theorem of Arason-Pfister, cf. [Lam 73, X.3.4]. The lemma follows by induction on the number of indices. □

Every Albert algebra A has a trace $\mathrm{Tr}\colon A \to k$ and a quadratic form q_A defined by $q_A(x) = \mathrm{Tr}(x^2)/2$. For the reduced Albert algebra J defined in 22.1, we have

$$q_J \cong \langle 2,2,2\rangle \oplus (q_3 \otimes \langle -\alpha_1, -\alpha_2, \alpha_1\alpha_2\rangle)\,, \tag{22.3}$$

cf. [SV 00, p. 118]. This quadratic form determines the isomorphism class of J by [SV 00, 5.8.1].

THEOREM 22.4. [Se 95, Th. 10] *Let A be an Albert algebra over k. There exist 3- and 5-Pfister forms q_3 and q_5 over k such that*

$$q_A \oplus q_3 \cong \langle 2,2,2\rangle \oplus q_5.$$

Moreover, this property characterizes q_3 and q_5 up to isomorphism and q_5 is divisible by q_3 (i.e., q_5 is isomorphic to a tensor product of q_3 with a 2-Pfister form).

PROOF. First suppose that A is reduced and q_3, q_5 are as in Example 22.1. By (22.3), the forms q_3 and $q_5 = q_2q_3$ satisfy the conditions of the theorem. If two other forms q_3' and q_5' also satisfy these conditions, then by (22.3) we have

$$\langle 2,2,2\rangle \oplus (q_5 - q_3) = \langle 2,2,2\rangle \oplus (q_5' - q_3')$$

in the Witt ring. Lemma 22.2 implies that $q_3 = q_3'$ and $q_5 = q_5'$, which finishes the case where A is reduced.

If the algebra A is not reduced, there is a cubic extension k' of k such that $A \otimes_k k'$ is reduced [SV 00, §6.1]. A descent property for Pfister forms due to Rost [Rost 99, §3] transfers the result from $A \otimes_k k'$ down to A. □

For an alternative approach, see [PR 95].

Define invariants f_3, $f_5 \in \mathrm{Inv}(\mathrm{Alb}, \mathbf{Z}/2\mathbf{Z})$ by setting $f_i = e_i(q_i)$ for $i = 3, 5$, where q_i is the Pfister form associated to an Albert algebra by the preceding theorem and $e_i \in \mathrm{Inv}(\mathrm{Pfister}_i, \mathbf{Z}/2\mathbf{Z})$ is as in §18.

THEOREM 22.5. $\mathrm{Inv}_{k_0}(\mathrm{Alb}, \mathbf{Z}/2\mathbf{Z})$ *is a free $H(k_0)$-module with basis $\{1, f_3, f_5\}$.*

PROOF. We have a morphism of functors $\mathrm{Pfister}_3 \times \mathrm{Pfister}_2 \to \mathrm{Alb}$ given by sending (q_3, q_2) to the reduced algebra constructed in Example 22.1. In this manner, a defines an invariant a_{Pf} of $\mathrm{Pfister}_3 \times \mathrm{Pfister}_2$. Recall that $H^1(k, \mathbf{Z}/2\mathbf{Z}) = k^*/k^{*2}$, and so there is a morphism of functors $H^1(*, \mathbf{Z}/2\mathbf{Z}) \to \mathrm{Pfister}_1$ given by $(\alpha) \mapsto \langle 1, -\alpha\rangle$. Similar statements hold for direct sums of n copies of $\mathbf{Z}/2\mathbf{Z}$ and $\mathrm{Pfister}_n$. Consequently, for $G = \mathbf{Z}/2\mathbf{Z} \times \cdots \times \mathbf{Z}/2\mathbf{Z}$ (5 copies), we have morphisms of functors $H^1(*, G) \to \mathrm{Pfister}_3 \times \mathrm{Pfister}_2 \to \mathrm{Alb}$. Any invariant $a \in \mathrm{Inv}(\mathrm{Alb}, \mathbf{Z}/2\mathbf{Z})$ restricts to an invariant on $H^1(*, G)$, hence is of the form described in 16.4. The arguments in the proof of 18.1 show that

$$a_{\mathrm{Pf}}(q_3, q_2) = \lambda_0 + \lambda_2 \cdot e_2(q_2) + \lambda_3 \cdot e_3(q_3) + \lambda_{23} \cdot e_3(q_3) \cdot e_2(q_2) \tag{22.6}$$

where the λ's are in $H(k_0)$.

The invariant a_{Pf} in turn defines an invariant a_2 in $\mathrm{Inv}_{k_0}(\mathrm{Pfister}_2, \mathbf{Z}/2\mathbf{Z})$ by setting

$$a_2(q_2) = a_{\mathrm{Pf}}(\mathrm{hyp}_3, q_2)$$

for hyp_3 the hyperbolic 3-Pfister form, which is a tensor product of three copies of $\langle 1, -1\rangle$. Since $e_3(\mathrm{hyp}_3)$ is 0, this invariant is given by

$$a_2(q_2) = \lambda_0 + \lambda_2 \cdot e_2(q_2).$$

The Albert algebra constructed from hyp_3 and any 2-Pfister form q_2 is well-known to be split. Thus a_2 is constant, and since $e_2(\mathrm{hyp}_2) = 0$ we have

$$0 = a_2(q_2) - a_2(\mathrm{hyp}_2) = \lambda_2 \cdot e_2(q_2).$$

By 18.1, this implies $\lambda_2 = 0$. Thus (22.6) may be rewritten as

$$a_{\mathrm{Pf}}(q_3, q_2) = \lambda_0 + \lambda_3 \cdot e_3(q_3) + \lambda_{23} \cdot e_3(q_3) \cdot e_2(q_2). \tag{22.7}$$

The invariant a is determined by its values on the reduced Albert algebras. For, let a' be an invariant in $\mathrm{Inv}_{k_0}(\mathrm{Alb}, \mathbf{Z}/2\mathbf{Z})$ which agrees with a on reduced algebras. Let A be a nonreduced Albert algebra over some extension k of k_0. There is a cubic extension k' of k over which A is reduced [SV 00, §6.1]. So $\mathrm{Res}_k^{k'}(a(A)) = a(A \otimes_k k')$ is equal to $a'(A \otimes_k k') = \mathrm{Res}_k^{k'} a'(A)$. Since $[k' : k]$ is odd, the map $H(k) \to H(k')$ is injective, hence $a(A)$ is equal to $a'(A)$.

Since a agrees with $\lambda_0 + \lambda_3 \cdot f_3 + \lambda_{23} \cdot f_5$ for reduced Albert algebras, the two invariants are the same. Thus the set $\{1, f_3, f_5\}$ spans $\mathrm{Inv}(\mathrm{Alb}, \mathbf{Z}/2\mathbf{Z})$.

Now suppose that $a = \lambda_0 + \lambda_3 \cdot f_3 + \lambda_{23} \cdot f_5$ is 0. Then $a_{\mathrm{Pf}} = 0$. In particular, λ_0 is 0. We thus get an invariant of $\mathrm{Pfister}_3$ by

$$q_3 \mapsto a_{\mathrm{Pf}}(q_3, \mathrm{hyp}_2) = \lambda_3 \cdot e_3(q_3).$$

Since this invariant is 0, we have $\lambda_3 = 0$ by 18.1.

The tensor product defines a morphism

$$\mathrm{Pfister}_3 \times \mathrm{Pfister}_2 \to \mathrm{Pfister}_5\,.$$

For every extension k of k_0, this is surjective. Hence it induces an injective map

$$\mathrm{Inv}_{k_0}(\mathrm{Pfister}_5, \mathbf{Z}/2\mathbf{Z}) \to \mathrm{Inv}_{k_0}(\mathrm{Pfister}_3 \times \mathrm{Pfister}_2, \mathbf{Z}/2\mathbf{Z}).$$

The invariant a_{Pf} is given by

$$a_{\mathrm{Pf}}(q_3, q_2) = \lambda_{23} \cdot e_2(q_2) \cdot e_3(q_3),$$

that is, it is the image of $\lambda_{23} \cdot e_5$. Since this is 0 by hypothesis, we have $\lambda_{23} = 0$ by 18.1. Thus all of the coefficients specifying a are 0. This shows that the set $\{1, f_3, f_5\}$ is linearly independent. □

For invariants mod 3 of Alb, see [Rost 91] and [KMRT 98, Chap. IX].

EXERCISE 22.8 (cup products of invariants of F_4). Show that the ring structure of $\mathrm{Inv}(F_4, \mathbf{Z}/2\mathbf{Z})$ (with product the cup product) is given by the formulas

$$f_3 \cdot f_3 = (-1) \cdot (-1) \cdot (-1) \cdot f_3$$
$$f_3 \cdot f_5 = (-1) \cdot (-1) \cdot (-1) \cdot f_5$$
$$f_5 \cdot f_5 = (-1) \cdot (-1) \cdot (-1) \cdot (-1) \cdot (-1) \cdot f_5.$$

[Hint: Use the fact that, if A is a Jordan algebra over k, there are five elements $x_1, \ldots,$ x_5 of k^* such that $f_3(A) = (x_1) \cdot (x_2) \cdot (x_3)$ and $f_5(A) = (x_1) \cdot (x_2) \cdot (x_3) \cdot (x_4) \cdot (x_5)$.]

EXERCISE 22.9 (invariants of E_6). Let E_6 be a split group of type E_6, either simply connected or adjoint. There are standard embeddings $G_2 \to F_4 \to E_6$. Show that the composite map

$$e\colon \mathrm{Torsors}_{G_2} \to \mathrm{Torsors}_{E_6}$$

is "almost" a bijection, in the following sense:

(1) e is injective;
(2) for every k-E_6-torsor x, there is an odd-degree extension k'/k over which x belongs to the image of e.

Prove also that the restriction map

$$\mathrm{Inv}(E_6, \mathbf{Z}/2\mathbf{Z}) \to \mathrm{Inv}(G_2, \mathbf{Z}/2\mathbf{Z})$$

is an *isomorphism.* In particular, $\mathrm{Inv}(E_6, \mathbf{Z}/2\mathbf{Z})$ has basis $\{1, f_3\}$ over $H(k_0)$, where f_3 is the mod 2 component of the Rost invariant in the case where E_6 is simply connected.

22.10. TESTING INVARIANTS ON SMALL SUBGROUPS? The reader may have observed that for several "large" groups G and primes p, one can find a "small" (e.g., finite abelian) subgroup H of G such that the restriction map

$$\mathrm{Res} : \mathrm{Inv}(G, C) \to \mathrm{Inv}(H, C)$$

is injective, for every C which is an abelian p-group. For $p = 2$, this is the case for the following groups G, with H abelian of type $(2, \dots, 2)$:

G	rank H	reference
O_n	n	elementary
SO_{n-1}	$n-1$	elementary
G_2	3	18.4
split F_4	5	22.5
split E_6	3	22.9
S_n	$[n/2]$	24.11

When $p > 2$, there are similar examples with H abelian of type $(p, \dots, p)$, for instance:

G	p	H
PGL_p	p	$\mathbf{Z}/p\mathbf{Z} \times \boldsymbol{\mu}_p$
split F_4	3	$\mathbf{Z}/3\mathbf{Z} \times \boldsymbol{\mu}_3 \times \boldsymbol{\mu}_3$
split simply connected E_6	3	$\mathbf{Z}/3\mathbf{Z} \times \boldsymbol{\mu}_3 \times \boldsymbol{\mu}_3 \times \boldsymbol{\mu}_3$
split E_7	3	$\mathbf{Z}/3\mathbf{Z} \times \boldsymbol{\mu}_3 \times \boldsymbol{\mu}_3 \times \boldsymbol{\mu}_3$
split E_8	5	$\mathbf{Z}/5\mathbf{Z} \times \boldsymbol{\mu}_5 \times \boldsymbol{\mu}_5$

[The first two are not difficult to prove, by methods similar to those used in the present chapter. The surjectivity of $H^1(*, F_4 \times \boldsymbol{\mu}_3) \to H^1(*, E_6)$, see e.g. [Ga 01, §3], gives the third. The last one follows from [Ch 94].]

Note that in each case, H is one of the subgroups used by Reichstein in [Re 00] for giving a lower bound of the essential dimension of G. It would be interesting to have more examples of such pairs (G, H); one should not expect, however, that one subgroup H would always be enough, nor that H would always be elementary abelian.

23. Cohomological invariants mod 2^m

Let C be a finite Γ_{k_0}-module whose order is a power of 2. In this section, we compute $\mathrm{Inv}(\mathrm{Quad}_n, C)$.

Let us say that a class $z \in H^i(k, \mathbf{Z}/2\mathbf{Z})$ is *liftable* if, for every $m \geq 1$, z belongs to the image of the natural map $H^i(k, \mathbf{Z}/2^m\mathbf{Z}(i)) \to H^i(k, \mathbf{Z}/2\mathbf{Z})$. Recall that "$(i)$" denotes the i-th Tate twist as in 7.8. Note that:

- If $i = 1$, every z is liftable, since $\mathbf{Z}/2^m\mathbf{Z}(i)$ is just the Galois module of 2^m-th roots of unity so that the map

$$H^1(k, \mathbf{Z}/2^m\mathbf{Z}(i)) \to H^1(k, \mathbf{Z}/2\mathbf{Z})$$

is none other than the projection $k^*/k^{*2^m} \to k^*/k^{*2}$.
- The cup product of two liftable elements is liftable.

- The sum of two liftable elements is liftable.

Hence any sum of cup products of elements of degree 1 — for example, any Stiefel-Whitney invariant of a quadratic form q — is liftable. (Milnor's conjecture implies that every z is liftable.)

Let $x \in H(k, C(-i))$ be such that $2x = 0$. Let $z \in H^i(k, \mathbf{Z}/2\mathbf{Z})$ be liftable. We define a cup product $z \bullet x$ in $H(k, C)$. To do so, choose m large enough so that C is killed by 2^{m-1}, and choose a lifting z_m of z, i.e., an element of $H^i(k, \mathbf{Z}/2^m\mathbf{Z}(i))$ whose image in $H^i(k, \mathbf{Z}/2\mathbf{Z})$ is z. There is an obvious pairing

$$\mathbf{Z}/2^m\mathbf{Z}(i) \times C(-i) \to C,$$

which allows us to define the (usual) cup product $z_m \cdot x$ in $H(k, C)$.

LEMMA 23.1. *$z_m \cdot x$ does not depend on the choice of lifting z_m, and is killed by 2.*

PROOF. A different lifting of z in $H^i(k, \mathbf{Z}/2^m\mathbf{Z}(i))$ differs from z_m by the image of a cohomology class $z' \in H^i(k, \mathbf{Z}/2^{m-1}\mathbf{Z}(i))$ by the natural map "multiplication by 2" : $\mathbf{Z}/2^{m-1}\mathbf{Z}(i) \to \mathbf{Z}/2^m\mathbf{Z}(i)$. This modifies $z_m \cdot x$ by $2z' \cdot x$, which is 0 since $2x = 0$ by assumption. □

We write $z \bullet x$ for $z_m \cdot x$, and call it the "cup product" of z and x. It belongs to $H(k, C)_2$, where the subscript 2 means the subgroup of elements killed by 2.

23.2. INVARIANTS OF $\mathbf{Z}/2\mathbf{Z}$. Choose $x \in H(k_0, C(-1))_2$. If k is an extension of k_0 and $z \in H^1(k, \mathbf{Z}/2\mathbf{Z})$, the element $z \bullet x$ of $H(k, C)$ is defined as above. The map $z \mapsto z \bullet x$ is an element of $\mathrm{Inv}_{k_0}(\mathbf{Z}/2\mathbf{Z}, C)$ which we denote by $\underline{\mathrm{id}} \bullet x$.

LEMMA. *Every normalized invariant in $\mathrm{Inv}_{k_0}(\mathbf{Z}/2\mathbf{Z}, C)$ can be written uniquely as $\underline{\mathrm{id}} \bullet x$ with $x \in H(k_0, C(-1))_2$.*

PROOF. Let $a \in \mathrm{Inv}(\mathbf{Z}/2\mathbf{Z}, C)$ be a normalized invariant. As in 16.2, $(t)_2$ is a versal $k_0(t)$-$\mathbf{Z}/2\mathbf{Z}$-torsor. The element $a((t)_2)$ is unramified outside $\{0, \infty\}$. Its residue at $t = 0$ is an element x_1 of $H(k_0, C(-1))$, and one sees as in 16.2 that the invariant is the cup product of $(t)_{2^m}$ with x_1. Since $(t)_2 = (t^3)_2$, we have $2(t)_{2^m} \cdot x_1 = 0$, hence $2x_1 = 0$ by 7.11. Therefore, $a((t)_2) = (t)_2 \bullet x_1$, as desired. □

23.3. INVARIANTS OF $G = \mathbf{Z}/2\mathbf{Z} \times \cdots \times \mathbf{Z}/2\mathbf{Z}$. Recall the invariant $a_I \in \mathrm{Inv}(G, \mathbf{Z}/2\mathbf{Z})$ from 16.4. For $z \in H^1(k, G)$, $a_I(z)$ is an element of $H^i(k, \mathbf{Z}/2\mathbf{Z})$ which is obviously liftable. For $x \in H(k_0, C(-|I|))_2$, we write $a_I \bullet x$ for the invariant of G given by $z \mapsto a_I(z) \bullet x$.

LEMMA. *Every normalized invariant of $G = \mathbf{Z}/2\mathbf{Z} \times \cdots \times \mathbf{Z}/2\mathbf{Z}$ (n times) can be written uniquely as*

$$\sum_{\substack{I \subset [1,n] \\ I \neq \emptyset}} a_I \bullet x_I \quad \textit{where } x_I \in H(k_0, C(-|I|))_2.$$

PROOF. By induction from 23.2, as for Theorem 16.4. □

23.4. INVARIANTS OF Quad_n. Let q be a quadratic form, and let x be an element of $H(k, C(-i))$ with $2x = 0$ as above. Since $w_i(q)$ is liftable, the cup product $w_i(q) \bullet x$ is a well-defined element of $H(k, C)_2$.

This gives a method for constructing an element of $\mathrm{Inv}_{k_0}(\mathrm{Quad}_n, C)$: start with any x_i in $H(k_0, C(-i))_2$, and take the invariant

$$q \mapsto w_i(q) \bullet x_i.$$

We denote this invariant by $w_i \bullet x_i$. Note that it takes values in the subgroup $H(k, C)_2$ of elements killed by 2.

THEOREM 23.5. *Any normalized invariant* $\mathrm{Quad}_n \to H(*, C)$ *can be written uniquely as*

$$\sum_{i=1}^{n} w_i \bullet x_i \quad \textit{with } x_i \in H(k_0, C(-i))_2.$$

PROOF. The result follows from 23.3 by the same method as 17.3. □

COROLLARY 23.6. *If q, $q' \in \mathrm{Quad}_n(k)$ have the same Stiefel-Whitney invariants, one has $a(q) = a(q')$ for every $a \in \mathrm{Inv}(\mathrm{Quad}_n, C)$.*

This "explains" why the Arason invariant e_3 is not the restriction of an invariant of Quad_n, cf. [Ar 75, last 3 lines]: Suppose that -1 is a square in k_0 and take two 3-Pfister forms q, q' over k/k_0. One checks that $w_i(q)$ and $w_i(q')$ vanish for $i \geq 1$. If e_3 were a linear combination of the w_i, we would have $e_3(q) = e_3(q')$, which is not always true.

EXERCISE 23.7. Let $a \in \mathrm{Inv}(\mathrm{Quad}_n, C)$ be a normalized invariant. Show that $2a = 0$. [Hint: use Th. 23.5 or give a direct argument as in the proof of Th. 24.12 below.]

23.8. RELATION WITH MILNOR K-THEORY. In order to use Milnor's notation, we write F for the ground field instead of k. The i-th Milnor K-group of F will be denoted by $K_i(F)$. Let C be a finite Γ_F-module of order prime to the characteristic. There exists a natural pairing

$$c \colon K_i(F) \times H(F, C) \to H(F, C(i))$$

characterized by the property that, if $x = (x_1, \ldots, x_i)$ is a decomposable element of $K_i(F)$ and z belongs to $H(F, C)$, then $c(x, z)$ is the cup product $(x_1) \cdots (x_i) \cdot z$. The fact that c exists is clear when $i = 1$; the crucial case $i = 2$ is a theorem of Tate [Ta 76, Th. 3.1]; the case $i > 2$ follows from the case $i = 2$ and the definition of the Milnor K-groups.

It is natural to see c as a kind of cup product and to denote it simply by $(x, z) \mapsto x \cdot z$.

Consider now the Milnor k-group $k_i(F) = K_i(F)/2K_i(F)$. Call $H(F, C)_2$ the subgroup of $H(F, C)$ killed by 2. It is clear that, if z belongs to that group, then $x \cdot z$ depends only on the image of x in $k_i(F)$. Hence we get a pairing

$$k_i(F) \times H(F, C)_2 \to H(F, C(i))_2.$$

Let now q be a quadratic form over F. By [Mi 70, §3], the Stiefel-Whitney class $w_i(q)$ may be defined in $k_i(F)$. Hence we may use it to map $H(F, C)_2$ in $H(F, C(i))_2$, hence also to map $H(F, C(-i))_2$ in $H(F, C)_2$. We find the same map as the one defined above.

CHAPTER VII

Cohomological invariants of étale algebras

24. Properties of cohomological invariants of Et_n

As before, we fix our functor H to be $k \mapsto H(k, C)$, where C is a finite Γ_{k_0}-module of order prime to the characteristic.

24.1. UNRAMIFIED ÉTALE ALGEBRAS. Let K be a field complete with respect to a discrete valuation v. We keep the notation of §7 and §11: R is the valuation ring of v (i.e., the set of $x \in K$ with $v(x) \geq 0$); k is the residue field.

Let E be an étale algebra of rank n over K, and let $\varphi_E : \Gamma_K \to S_n$ be the corresponding homomorphism (well-defined up to conjugation), cf. 3.2. If we view φ_E as a cocycle, its class in $H^1(K, S_n) = \mathrm{Et}_n(K)$ will be denoted by (E).

PROPOSITION 24.2. *The following properties are equivalent:*

(1) $\varphi_E(I_K) = 1$, *where I_K is the inertia subgroup of Γ_K, cf.* 7.1.
(2) *E is a product of fields which are unramified extensions of K.*
(3) *The class (E) in $H^1(K, S_n)$ belongs to the image of the canonical injection $i\colon H^1(k, S_n) \to H^1(K, S_n)$, cf.* 11.1.
(4) *There is a K-basis $(e_1, \ldots, e_n)$ of E such that the matrix $(\mathrm{Tr}(e_i.e_j))$ belongs to $\mathbf{GL}_n(R)$.*

PROOF. The equivalences (1) $\iff$ (2) $\iff$ (3) $\iff$ (4) are clear. □

DEFINITION 24.3. An étale algebra E is called *unramified* if it has properties (1) through (4) of 24.2.

EXAMPLE 24.4. If $\mathrm{char}(k) \neq 2$, a quadratic algebra $K[X]/(X^2 - a)$ is unramified if and only if $v(a) \equiv 0 \pmod 2$.

Let $a \in \mathrm{Inv}(\mathrm{Et}_n, C)$ be a cohomological invariant, over a ground field k_0 which is contained in R (and hence also in k and in K).

PROPOSITION 24.5. *If E is unramified, then the residue of $a(E)$ is* 0.

PROOF. This follows from 11.7. □

24.6. A VERSAL TORSOR. Let $K = k_0(c_1, \ldots, c_n)$ where the c_i are indeterminates, and set $E^{\mathrm{gen}} = K[t]/(t^n + c_1 t^{n-1} + \cdots + c_n)$ as in 5.6.2. For k infinite, every étale k-algebra can be generated by one element [Bou 59, V.7.7, Prop. 7] and it is easily seen that the generators are Zariski-dense, so that $E^{\mathrm{gen}} \in \mathrm{Et}_n(K)$ corresponds to a versal torsor in $H^1(K, S_n)$. The fact that this is versal may also be viewed as a special case of 5.4, applied to the natural embedding of S_n in $\mathbf{GL}_n$. (This will be used in part (2) of the proof of Prop. 24.6 below.)

Let D be an irreducible divisor of $\mathrm{Aff}^n = \mathrm{Spec}\, k_0[c_1, \ldots, c_n]$. Such a divisor defines a discrete valuation v_D of $K = k_0(c_1, \ldots, c_n)$; we write K_D for the completion of K with respect to v_D. Its residue field is the function field $k_0(D)$ of D. Let E_D^{gen} denote the image of E^{gen} under the restriction map $H^1(K, S_n) \to H^1(K_D, S_n)$, or equivalently under the map $\text{Et}_n(K) \to \text{Et}_n(K_D)$.

Let Discr be the discriminant of the polynomial $t^n + c_1t^{n-1} + \cdots + c_n$. When $n = 1$, we have Discr $= 1$. When $n \geq 2$, Discr is an irreducible polynomial in c_1, ..., c_n, except when $\text{char}(k) = 2$, in which case it is the square of an irreducible polynomial. It defines an irreducible divisor Δ in Aff^n.

PROPOSITION. (1) *If $D \neq \Delta$, the K_D-algebra E_D^{gen} is unramified* (in the sense of 24.3).

(2) *If $D = \Delta$, then E_D^{gen} is the product of a quadratic algebra and an unramified algebra of rank $n-2$.*

PROOF. (1): For a basis $(e_1, \ldots, e_n)$ of E^{gen}, choose $(1, t, \ldots, t^{n-1})$. We have $\text{Tr}(e_i.e_j) \in k_0[c_1, \ldots, c_n]$ and the determinant of the matrix $M = (\text{Tr}(e_i.e_j))$ is equal to $\text{Discr}(c_1, \ldots, c_n)$. Hence M belongs to $\mathbf{GL}_n(R_D)$, where R_D is the valuation ring of v_D. Criterion (4) of 24.2 then shows that E_D^{gen} is unramified.

(Note that the "geometric" proof of (2) given below also provides a proof of (1).)

(2): Let Aff_x^n denote $\text{Spec}(k_0[x_1, \ldots, x_n])$. Write Aff_c^n for the affine space with coordinates $c_1, \ldots, c_n$, which we interpret as symmetric functions in $x_1, \ldots, x_n$. The group S_n acts on Aff_x^n with quotient Aff_c^n, and it acts freely outside the "bad locus" (= ramification locus) consisting of the union of the hyperplanes H_{ij} defined by the equations $x_i = x_j$ for $i \neq j$. The map $\text{Aff}_x^n \to \text{Aff}_c^n$ is unramified outside of the image Δ of the bad locus, and completing K to obtain K_Δ corresponds to completing Aff_c^n at the generic point of the bad locus.

The algebra $E_\Delta^{\text{gen}} = E^{\text{gen}} \otimes_K K_D$ is described by a map $\Gamma_{K_\Delta} \to S_n$ and the geometric description shows that the image of the inertia group is the subgroup of S_n generated by a transposition $(i\,j)$. Since the inertia group is normal in Γ_{K_Δ}, the algebra E_Δ^{gen} is isomorphic to the product of a ramified quadratic K_Δ-algebra corresponding to $\{i, j\}$ and an unramified K_Δ-algebra of rank $n-2$ corresponding to the complement of $\{i, j\}$ in $\{1, 2, \ldots, n\}$. □

EXERCISE 24.7. In case (2) of Prop. 24.6, show that both factors of E_Δ^{gen} are fields, if $n \geq 3$.

24.8. SPLITTING PRINCIPLE. We call an étale algebra *multiquadratic* if it is decomposable as a product of étale algebras of rank 1 or 2.

THEOREM 24.9. *If $a \in \text{Inv}_{k_0}(\text{Et}_n, C)$ satisfies $a(E) = 0$ for every multiquadratic $E \in \text{Et}_n(k)$ (over every extension k of k_0), then $a = 0$.*

PROOF. By induction on n. The result is clear when $n = 1$ or 2, so we assume that $n \geq 3$.

Suppose that $E \in \text{Et}_n(k)$ can be written as $E = E' \times E''$ where E' has rank $i = 1$ or 2. Define an invariant $\text{Et}_{n-i/k} \to H$ by $F \mapsto a(E' \times F)$ where F runs through the $(n-i)$-étale algebras over the extensions of k. This invariant vanishes for multiquadratic extensions, so by induction it is zero. Hence in this case we have $a(E) = 0$.

Let E^{gen}, Δ, etc. be as in 24.6. We consider the residues of $\alpha = a(E^{\text{gen}})$ corresponding to irreducible divisors of the affine space Aff^n.

Let D be such an irreducible divisor. If $D \neq \Delta$, then by 24.4 and 24.6.1, the image of α in $H(K_D, C)$ belongs to $H(k(D), C)$. Thus α has residue 0 at D. For $D = \Delta$, E^{gen}_Δ is isomorphic to $E' \times E''$ with E' of rank 2. As shown above, this implies that the image of α in $H(K_\Delta, C)$ is 0. Hence the residue of α with respect to the discrete valuation corresponding to Δ is 0.

Since α has residue 0 with respect to all discrete valuations of K/k_0 which correspond to irreducible hypersurfaces, α belongs to $H(k_0, C)$ by 10.1. That is, a and the constant invariant defined by α agree on E^{gen}. Hence these invariants are equal since E^{gen} is versal (cf. 12.3). Since a vanishes for multiquadratic algebras, a is identically 0. □

REMARK 24.10. Let H be the subgroup of S_n generated by the transpositions $(2j-1, 2j)$, for $j = 1, \ldots, m = [n/2]$. It is an elementary abelian group of type $(2, \ldots, 2)$ and rank m. Theorem 24.9 can be restated in the following form:

THEOREM 24.11. *The restriction map* (cf. §13)

$$\mathrm{Res}^H_{S_n} : \mathrm{Inv}(S_n, C) \to \mathrm{Inv}(H, C)$$

is injective.

(Recall that $\mathrm{Inv}(S_n, C) = \mathrm{Inv}(\mathrm{Et}_n, C)$, see 3.2.)

PROOF. Indeed, H-torsors correspond to sequences $(E_1, \ldots, E_m)$ of m quadratic étale algebras. Hence, if an invariant a is such that $\mathrm{Res}(a) = 0$, one has

$$a(E_1 \times \cdots \times E_m) = 0 \quad (\text{if } n \text{ is even, i.e., } n = 2m)$$
$$a(E_1 \times \cdots \times E_m \times k) = 0 \quad (\text{if } n \text{ is odd, i.e., } n = 2m+1)$$

for all possible choices of $E_1, \ldots, E_m$ over every k/k_0. By the above theorem, this implies $a = 0$. □

24.12. Recall that $a \in \mathrm{Inv}_{k_0}(\mathrm{Et}_n, C)$ is *normalized* in the sense of 4.5 if $a(E^{\text{split}}) = 0$ in $H(k_0, C)$ for $E^{\text{split}} = k_0 \times \cdots \times k_0$.

THEOREM. *If* $a \in \mathrm{Inv}(\mathrm{Et}_n, C)$ *is normalized, then* $2a = 0$.

PROOF. By Theorem 24.9, it is enough to prove that $2a(E) = 0$ for $E = E' \times E''$ where E' has rank 1 or 2.

Set $E_1 = k \times E''$ or $k \times k \times E''$ as needed so that it has the same rank as E. The same construction as at the start of the proof of Theorem 24.9 combined with induction on n shows that $2a(E_1) = 0$.

There is an extension k' of k of degree at most 2 such that E and E_1 become isomorphic over k'. Hence $a(E)$ and $a(E_1)$ have the same image in $H(k', C)$. By applying $\mathrm{Cor}^{k'}_k$ we get $[k' : k] \cdot a(E) = [k' : k] \cdot a(E_1)$, hence $2a(E) = 2a(E_1) = 0$. □

25. Cohomological invariants of Et_n

From now on, until §33, the characteristic is $\neq 2$.

In this section, we fix $C = \mathbf{Z}/2\mathbf{Z}$ and construct invariants

$$\mathrm{Et}_n \to H(k, \mathbf{Z}/2\mathbf{Z}) = H(k).$$

25.1. FIRST CONSTRUCTION OF INVARIANTS: THE GALOIS STIEFEL-WHITNEY CLASSES. Each $u \in H(S_n, \mathbf{Z}/2\mathbf{Z})$ defines a mod 2 cohomological invariant of Et_n

by pulling back the cohomology of S_n as in 4.2: Every $E \in \mathrm{Et}_n(k)$ defines a map $\varphi_E : \Gamma_k \to S_n$ as in 3.2, and we set $a_u(E) = \varphi_E^*(u)$.

We have obvious embeddings $S_n \hookrightarrow \mathbf{O}_n \hookrightarrow \mathbf{GL}_n$ where $\mathbf{O}_n$ and $\mathbf{GL}_n$ denote the groups over $\mathbf{R}$ endowed with the usual topology. Let $B_{\mathbf{O}_n}$ and B_{S_n} be the "classifying spaces" of $\mathbf{O}_n$ and S_n, cf. [Bor 55, §8]. The map $B_{S_n} \to B_{\mathbf{O}_n}$ defines a map of singular cohomology groups

$$H(B_{\mathbf{O}_n}, \mathbf{Z}/2\mathbf{Z}) \to H(B_{S_n}, \mathbf{Z}/2\mathbf{Z}).$$

Since S_n is discrete, the cohomology of S_n (in the algebraic sense of the word) is the same as the (topological) cohomology of the classifying space B_{S_n}; that is, we have the (sometimes puzzling) identity $H(B_{S_n}, \mathbf{Z}/2\mathbf{Z}) = H(S_n, \mathbf{Z}/2\mathbf{Z})$. Thus any element of $H(B_{\mathbf{O}_n}, \mathbf{Z}/2\mathbf{Z})$ defines a corresponding element of $H(B_{S_n}, \mathbf{Z}/2\mathbf{Z}) = H(S_n, \mathbf{Z}/2\mathbf{Z})$. In particular, the Stiefel-Whitney classes $w_1, \ldots, w_n$ of $H(B_{\mathbf{O}_n}, \mathbf{Z}/2\mathbf{Z})$ (see e.g. [Bor 55, §10]) define elements of $H(S_n, \mathbf{Z}/2\mathbf{Z})$ which we still denote by w_1, $\ldots$, w_n. The associated elements of $\mathrm{Inv}(\mathrm{Et}_n, \mathbf{Z}/2\mathbf{Z})$ are denoted by w_i^{gal}.

25.2. Total Galois Stiefel-Whitney class. We adopt the convention that $w_0^{\mathrm{gal}} = 1$ and that $w_i^{\mathrm{gal}} = 0$ when $i < 0$ or $i > n$, and we set

$$w^{\mathrm{gal}}(E) = \sum_i w_i^{\mathrm{gal}}(E).$$

As in 17.2, we have

$$w^{\mathrm{gal}}(k[x]/(x^2 - \alpha)) = 1 + (\alpha)$$
$$w^{\mathrm{gal}}(E \times E') = w^{\mathrm{gal}}(E) \cdot w^{\mathrm{gal}}(E')$$

Example 25.3. $w_1 \in H^1(S_n, \mathbf{Z}/2\mathbf{Z}) = \mathrm{Hom}(S_n, \mathbf{Z}/2\mathbf{Z})$ is the sign homomorphism, and w_1^{gal} is the discriminant.

$w_2 \in H^2(S_n, \mathbf{Z}/2\mathbf{Z})$ determines a central extension $\widetilde{S_n}$ of S_n by $\mathbf{Z}/2\mathbf{Z}$ described by Schur [Schur 11]: There is an exact sequence

$$1 \to \{\pm 1\} \to \widetilde{S_n} \to S_n \to 1$$

characterized by the fact that a transposition in S_n lifts to an element of order 2 of $\widetilde{S_n}$, but a product of 2 disjoint transpositions lifts to an element of order 4. So w_2 is not trivial if $n \geq 4$.

Example 25.4. Let $E = E_1 \times \cdots \times E_m$ where $E_j = k[x]/(x^2 - x_j)$ for some $x_j \in k^*/k^{*2}$. The properties in 25.2 give

$$w^{\mathrm{gal}}(E) = \prod_j (1 + (x_j)).$$

25.5. Description of $\mathrm{Inv}(\mathrm{Et}_n, \mathbf{Z}/2\mathbf{Z})$ in terms of the w_i^{gal}. Let $m = [n/2]$, the integral part of $n/2$, and let $H = \mathbf{Z}/2\mathbf{Z} \times \cdots \times \mathbf{Z}/2\mathbf{Z}$ (m copies). We identify H with the subgroup of S_n generated by the m transpositions $(1\,2)$, $(3\,4)$, ... By Th. 24.11 the restriction map

$$\mathrm{Inv}_{k_0}(\mathrm{Et}_n, \mathbf{Z}/2\mathbf{Z}) = \mathrm{Inv}_{k_0}(S_n, \mathbf{Z}/2\mathbf{Z}) \to \mathrm{Inv}_{k_0}(H, \mathbf{Z}/2\mathbf{Z})$$

is injective. Moreover, by 13.2, its image lies in the subgroup $\mathrm{Inv}_{k_0}(H, \mathbf{Z}/2\mathbf{Z})^{S_m}$ of $\mathrm{Inv}_{k_0}(H, \mathbf{Z}/2\mathbf{Z})$ fixed by the action of the symmetric group S_m (acting on H by permuting the factors).

THEOREM 25.6. (1) *The map* $\mathrm{Res}\colon \mathrm{Inv}_{k_0}(\mathrm{Et}_n, \mathbf{Z}/2\mathbf{Z}) \to \mathrm{Inv}_{k_0}(H, \mathbf{Z}/2\mathbf{Z})^{S_m}$ *is an isomorphism.*

(2) *The* $H(k_0)$*-module* $\mathrm{Inv}_{k_0}(\mathrm{Et}_n, \mathbf{Z}/2\mathbf{Z})$ *is free with basis* 1, $w_1^{\mathrm{gal}}, \ldots, w_m^{\mathrm{gal}}$, *where* $m = [n/2]$.

(3) *One has* $w_i^{\mathrm{gal}} = 0$ *for* $i > m$.

PROOF. The argument in the proof of Theorem 17.3 shows that $\mathrm{Inv}(H, \mathbf{Z}/2\mathbf{Z})^{S_m}$ is a free $H(k_0, \mathbf{Z}/2\mathbf{Z})$-module with basis consisting of the elementary symmetric functions $\sum_{i_1<\cdots<i_j}(x_{i_1})\cdots(x_{i_j})$ for $0 \le j \le m$. But such a symmetric function is the image of w_j^{gal} by Example 25.4. This shows that the map

$$\mathrm{Inv}(\mathrm{Et}_n, \mathbf{Z}/2\mathbf{Z}) \to \mathrm{Inv}(H, \mathbf{Z}/2\mathbf{Z})^{S_m}$$

is surjective. It is injective by Theorem 24.11, so it is an isomorphism. The w_i^{gal} are 0 for $i > m$ since this is true for multiquadratic algebras by 25.4. The theorem follows. □

25.7. SECOND CONSTRUCTION OF INVARIANTS: STIEFEL-WHITNEY CLASSES OF THE TRACE FORM. If $E \in \mathrm{Et}_n(k)$, its *trace form* q_E is the quadratic form on E defined by $q_E(x) = \mathrm{Tr}_{E/k}(x^2)$. This gives a morphism $\mathrm{Et}_n \to \mathrm{Quad}_n$. The Stiefel-Whitney invariants of Quad_n (as in 17.1) thus define cohomological invariants w_i on Et_n by

$$w_i(E) = w_i(q_E). \tag{25.8}$$

EXAMPLE 25.9. Let E be the multiquadratic algebra from Example 25.4. The trace form of E is $q_E = q_{E_1} \oplus \cdots \oplus q_{E_m}$, and $q_{E_i} = \langle 2, 2x_i \rangle$.

THEOREM 25.10. [Kahn 84] *For* $E \in \mathrm{Et}_n(k)$,

$$w_i^{\mathrm{gal}}(E) = \begin{cases} w_i(q_E) & \text{if } i \text{ is odd} \\ w_i(q_E) + (2)\cdot w_{i-1}(q_E) & \text{if } i \text{ is even} \end{cases}$$

in $H(k, \mathbf{Z}/2\mathbf{Z})$.

PROOF. By Theorem 24.9 it is sufficient to prove the formula when E is multiquadratic. Let us assume that n is even. (The case where n is odd is analogous.) By Example 25.9,

$$q_E = \underbrace{\langle 2, \ldots, 2 \rangle}_{m \text{ times}} \oplus \langle 2x_1, \ldots, 2x_m \rangle.$$

Hence $q_E = \langle 2 \rangle q'_E$ where $q'_E = \langle 1, \ldots, 1 \rangle \oplus \langle x_1, \ldots, x_m \rangle$. It is clear that $w_i(q'_E) = w_i^{\mathrm{gal}}(E)$. The formula follows from Lemma 25.12 below, which relates $w_i(q'_E)$ with $w_i(q_E)$. □

EXERCISE 25.11. Show that Th. 25.10 can be restated in terms of total Stiefel-Whitney classes as

$$w^{\mathrm{gal}}(E) = w(q_E)\cdot(1 + (2)\cdot(d)),$$

where d is the discriminant of E, so that $(d) = w_1(q_E) = w_1^{\mathrm{gal}}(E)$. Show that this formula is equivalent to:

$$w(q_E) = w^{\mathrm{gal}}(E)\cdot(1 + (2)\cdot(d)).$$

LEMMA 25.12. *If q is a quadratic form of rank n, then*

$$w_i(\langle 2\rangle q) = \begin{cases} w_i(q) & \text{if } n \equiv i \pmod 2 \\ w_i(q) + (2)\cdot w_{i-1}(q) & \text{if } n \not\equiv i \pmod 2. \end{cases}$$

PROOF. First note that we have $(2)\cdot(2) = 0$ in $H(k, \mathbf{Z}/2\mathbf{Z})$ since the quadratic form $\langle 1, -2, -2\rangle$ represents 0.

Write $q = \langle \alpha_1, \ldots, \alpha_n\rangle$. Recall that $w_i(q)$ is the i-th elementary symmetric polynomial in the (α_j)'s. We may expand $w_i(\langle 2\rangle q)$ using $(2\alpha_j) = (2) + (\alpha_j)$. Since $(2)\cdot(2) = 0$, most of the terms occurring in the expansion are 0 and we are left with

$$w_i(\langle 2\rangle q) = w_i(q) + \ell\cdot(2)\cdot w_{i-1}(q).$$

where ℓ is the integer $n - i + 1$. □

Theorem 25.10 combined with Theorem 25.6 gives:

THEOREM 25.13. *Set $m = [n/2]$. Then $\mathrm{Inv}_{k_0}(\mathrm{Et}_n, \mathbf{Z}/2\mathbf{Z})$ is a free $H(k_0)$-module with basis $1, w_1, \ldots, w_m$. We have $w_i = 0$ for $i > m+1$, and w_{m+1} is equal to 0 if m is even and is equal to $(2)\cdot w_m$ if m is odd.*

REMARKS 25.14. (1) From a practical point of view, the $w_i(q_E)$ are very easy to compute. If, for instance, $E = k[x]/P(x)$, one can write q_E in terms of the coefficients of P, diagonalize it, and get the w_i. The same would not be true for the $w_i^{\mathrm{gal}}(E)$.

Indeed, the case $i = 2$ [Se 84] has been used very often to compute $w_2^{\mathrm{gal}}(E)$, see e.g. [Vi 85], [BLV 86], [Vi 88], [Cr 89], [Cr 90], [Bö 90], [BF 91], [Hä 92], [Mes 95], [Mes 98]. This invariant has a Galois-theoretic interpretation: By 25.3, it is the *obstruction* to lifting the homomorphism $\varphi_E\colon \Gamma_k \to S_n$ as in 3.2 to a homomorphism $\Gamma_k \to \widetilde{S_n}$.

(2) Note that, if n is even, the cohomological invariants *are the same for n and $n+1$*. This fits well with Theorem 31.23.

25.15. GENERALIZATION TO WEYL GROUPS. Here we assume for simplicity that k_0 has characteristic 0. Let W be a Weyl group, i.e., a finite Coxeter group which stabilizes a lattice in $\mathbf{R}^n$ (i.e., a product of Weyl groups of type $A, \ldots, E_8$). The standard representation of W is defined over $\mathbf{Q}$, hence over k_0. One can thus make W act on Aff^n, with quotient isomorphic to Aff^n. Using this action (instead of the action of S_n), one can extend to W most of what has been proved above. In particular:

THEOREM. *An element of $\mathrm{Inv}(W, C)$ is 0 if its restriction to every abelian subgroup of W generated by reflections is 0.*

This implies that any normalized invariant is killed by 2.

25.16. INVARIANTS MOD 2^m. Now let C be a finite Γ_{k_0}-module whose order is a power of 2. For E an étale algebra, $w_i(E) \in H^i(k, \mathbf{Z}/2\mathbf{Z})$ is liftable in the sense of §23. For $x_i \in H(k_0, C(-i))_2$, we write $w_i \bullet x_i$ for the invariant

$$E \mapsto w_i(E) \bullet x_i \quad \text{in } \mathrm{Inv}(\mathrm{Et}_n, C),$$

where $\bullet$ is as defined in §23. By the same arguments as for $\mathrm{Inv}(\mathrm{Et}_n, \mathbf{Z}/2\mathbf{Z})$ above, we have:

THEOREM. *Every normalized invariant* $\mathrm{Et}_n \to H(*, C)$ *can be written uniquely as*

$$\sum_{i=1}^{[n/2]} w_i \bullet x_i \quad \textit{with } x_i \in H(k_0, C(-i))_2.$$

The same statement holds with w_i replaced with w_i^{gal}.

The theorem shows that *two étale algebras E and E' which have the same w_i's have the same cohomological invariants for any C*; in particular, this applies when $q_E \cong q_{E'}$.

EXERCISE 25.17. Let Γ be a profinite group and let X, X' be two finite Γ-sets (= sets with a continuous Γ-action). We say that X and X' are *weakly equivalent* (cf. [Se 71, §13.1, Exerc. 5]) if the corresponding complex linear representations of Γ are isomorphic, or, equivalently, if every $g \in \Gamma$ fixes the same number of points in X and in X'.

When $\Gamma = \Gamma_k$ and X, X' are of order n, the actions of Γ on X and X' give homomorphisms φ, $\varphi' \colon \Gamma \to S_n$ and thus define étale algebras E and E' over k. Suppose that $\mathrm{char}(k) \neq 2$ and the following holds:

If S is a 2-Sylow subgroup of Γ_k, then X and X' are weakly equivalent as S-sets.

Then:

(1) Show that q_E and $q_{E'}$ have the same Stiefel-Whitney classes.
[Hint: After replacing k by a tower of odd-degree extensions, we may assume that Γ is a pro-2-group, i.e., that $\Gamma = S$. Show that φ and φ' are then conjugate in $\mathbf{O}_n(\mathbf{R})$, and deduce that they give the same w_i; use 25.16.]
(2) Give an example where q_E and $q_{E'}$ are not isomorphic.
[Hint: Take $k = \mathbf{Q}(u, v)$, where u, v are indeterminates; write k_u, k_v, k_{uv} for the quadratic algebras defined by u, v, uv. Take $E = k_u \times k_v \times k_{uv}$ and $E' = k \times k \times (k_u \otimes k_v)$.]
(3) Assume that k contains an algebraic closure of the prime field. Show that q_E and $q_{E'}$ are isomorphic.

26. Applications to negligible cohomology

Let G be a finite group and M a G-module. An element $x \in H(G, M)$ is said to be *negligible* (cf. [BS 94] or [Se 94]) if, for every field k and every (continuous) homomorphism $\varphi \colon \Gamma_k \to G$, we have $\varphi^*(x) = 0$ in $H(\Gamma_k, M) = H(k, M)$.

(There is also a weaker notion of "geometrically negligible" studied by Saltman [Sal 95], where one asks for the same property, but only for fields k containing $\overline{\mathbf{Q}}$.)

PROPOSITION 26.1. *An element $x \in H(G, M)$ is negligible if $\varphi^*(x) = 0$ for every field k of characteristic* 0 *and every* $\varphi \colon \Gamma_k \to G$.

PROOF. Let x in $H^i(G, M)$ be negligible with respect to fields of characteristic 0. Consider a field k of characteristic $\neq 0$, and a homomorphism $f : \Gamma_k \to G$. We have to show that $f^*(x) = 0$ in $H^i(\Gamma_k, M)$. We may assume that k is perfect. Consider the ring of Witt vectors over k and its field of fractions K. We have a natural map $\Gamma_K \to \Gamma_k$ (see 7.1), hence f defines $f_0 : \Gamma_K \to G$. Since $\mathrm{char}\, K = 0$, we have $f_0^*(x) = 0$ in $H^i(\Gamma_K, M)$. But $\Gamma_K \to \Gamma_k$ splits by Lemma 7.6, hence the map $H^i(\Gamma_k, M) \to H^i(\Gamma_K, M)$ is injective, hence $f^*(x) = 0$. □

One can show that it is enough to check the condition $\varphi^*(x) = 0$ when k is of characteristic 0, and φ corresponds to a $\mathbf{Q}$-versal G-torsor.

EXAMPLE 26.2. Let $M = \mathbf{Z}/2\mathbf{Z}$ and fix $x, y \in H^1(G, \mathbf{Z}/2\mathbf{Z})$. If one writes X, Y for the images of x, y in $H^1(k, \mathbf{Z}/2\mathbf{Z})$, then one has $X \cdot X = X \cdot (-1)$, $Y \cdot Y = Y \cdot (-1)$, and hence

$$X \cdot X \cdot Y + X \cdot Y \cdot Y = 2X \cdot Y \cdot (-1) = 0.$$

That is, $x \cdot y \cdot (x + y)$ is negligible. It is nonzero for well-chosen G, x, and y, for example, when $G = \mathbf{Z}/2\mathbf{Z} \times \mathbf{Z}/2\mathbf{Z}$ and $\{x, y\}$ is a basis of $H^1(G, \mathbf{Z}/2\mathbf{Z})$.

Note that $x \cdot x$ is geometrically negligible but is not negligible in general.

THEOREM 26.3. *Let $G = S_n$ and assume that M is finite, with trivial action.*

(1) *An element x of $H^i(S_n, M)$ is negligible if and only if its restriction to every 2-elementary subgroup of S_n generated by transpositions is 0.*
(2) *For every $x \in H^i(S_n, M)$, $i > 0$, the cohomology class $2x$ is negligible.*
(3) *An element x of $H^i(S_n, \mathbf{Z}/2\mathbf{Z})$ is negligible if and only if its restriction to the subgroups of order ≤ 2 of S_n is 0.*

PROOF. Since S_n acts trivially on M, every element of $H^i(S_n, M)$ defines a cohomological invariant of Et_n as in 25.1. This gives a natural map

$$H^i(G, M) \to \mathrm{Inv}_{\mathbf{Q}}(S_n, M) = \mathrm{Inv}_{\mathbf{Q}}(\mathrm{Et}_n, M).$$

The kernel of this map clearly contains the subgroup $H^i_{\mathrm{negl}}(S_n, M)$ consisting of the negligible classes. Conversely, any element of the kernel is negligible for every field of characteristic 0, hence by Proposition 26.1 is negligible.

Property (1) is then a consequence of Theorem 24.9 and (2) follows from 24.12. Theorem 24.9 combined with the next lemma proves (3). □

LEMMA 26.4. [BS 94, Prop. 7.2.3] *Let H be an elementary group of type $(2, \ldots, 2)$, and let $\alpha \in H^i(H, \mathbf{Z}/2\mathbf{Z})$ be a cohomology class. The following are equivalent:*

(i) *α is negligible.*
(ii) *The restriction of α to every subgroup of order ≤ 2 of H is 0.*
(iii) *If x_λ is a basis of $H^1(H, \mathbf{Z}/2\mathbf{Z})$, α is in the ideal of the ring $H(H, \mathbf{Z}/2\mathbf{Z})$ generated by the $x_\lambda^2 x_\mu + x_\lambda x_\mu^2$.*

PROOF. (i) $\Rightarrow$ (ii) because 0 is the only negligible element of $H(G, \mathbf{Z}/2\mathbf{Z})$ if G has order 2 since such a G is isomorphic to $\mathrm{Gal}(\mathbf{C}/\mathbf{R})$.

(ii) $\Rightarrow$ (iii): The ring $H(H, \mathbf{Z}/2\mathbf{Z})$ is a polynomial ring over $\mathbb{F}_2$ in m variables x_λ (where $m = \operatorname{rank} H$), and α can be written as a homogeneous polynomial P of degree i in the x_λ. Assumption (ii) means that this polynomial vanishes whenever the coordinates are in $\mathbf{F}_2$. It is a simple exercise to show that this is equivalent to α being in the ideal generated by the $x_\lambda^2 x_\mu + x_\lambda x_\mu^2$.

(iii) $\Rightarrow$ (i): The element $x_\lambda^2 x_\mu + x_\lambda x_\mu^2$ is negligible as explained in Example 26.2. □

CHAPTER VIII

Witt invariants

Recall that the characteristic is assumed to be $\neq 2$.

27. Properties of the Witt ring

27.1. Let $WGr(k)$ be the Witt-Grothendieck ring of k. Recall that an element of $WGr(k)$ is a formal difference $q - q'$ where q and q' are nondegenerate quadratic forms over k and the sum and product are those induced by the orthogonal sum and tensor product of quadratic forms. The Witt ring $W(k)$ is the quotient of $WGr(k)$ by the ideal made up of integral multiples of $\mathrm{hyp}_2 = \langle 1, -1\rangle$. There is a short exact sequence

$$0 \to I \to W(k) \xrightarrow{\text{rank}} \mathbf{Z}/2\mathbf{Z} \to 0$$

where $I = I(k)$ is the "augmentation ideal" of $W(k)$. The diagram

$$\begin{array}{ccc} WGr(k) & \xrightarrow{\text{rank}} & \mathbf{Z} \\ \downarrow & & \downarrow \\ W(k) & \xrightarrow{\text{rank}} & \mathbf{Z}/2\mathbf{Z} \end{array}$$

is cartesian; i.e., an element of $WGr(k)$ may be viewed as a pair (q, n), with $n \in \mathbf{Z}$, $q \in W(k)$, and $n \equiv \mathrm{rank}(q) \pmod 2$.

Both W and WGr are functorial in k; we may take them as our "functor H". For most questions, $W(k)$ is more convenient to use. However, $WGr(k)$ has the advantage that it is a *λ-ring* in the sense of Grothendieck [SGA 6, Exp. V, §2]. If q is a quadratic form on a k-vector space V, let b be the unique symmetric bilinear form on V such that $q(x) = b(x, x)$ for every $x \in V$. Let $\lambda^p b$ be the corresponding symmetric form on $\wedge^p V$, cf. [Bou 59, IX.1.9]. The quadratic form on $\wedge^p V$ defined by $x \mapsto (\lambda^p b)(x, x)$ is denoted by $\lambda^p(q)$. If $q = \langle \alpha_1, \dots, \alpha_n\rangle$, we have $\lambda^p q = \sum_I \langle \alpha_I \rangle$, where I runs over the subsets of $[1, n]$ with $|I| = p$ and α_I is the product of the α_i for $i \in I$. The power series

$$\lambda_t(q) = \sum t^i \cdot \lambda^i(q) = 1 + t \cdot q + t^2 \cdot \lambda^2(q) + \cdots \quad \in WGr(k)[[t]]$$

is such that

$$\lambda_t(q + q') = \lambda_t(q) \cdot \lambda_t(q').$$

Thanks to this multiplicativity property, $\lambda_t(q)$ may be defined for every $q \in WGr(k)$, see [SGA 6, loc.cit.]. We thus have natural exterior power operations

$$\lambda^p \colon WGr(k) \to WGr(k).$$

EXERCISE 27.2. (1) Show that the two properties

$$\lambda_t(q+q') = \lambda_t(q)\cdot\lambda_t(q')$$
$$\lambda_t(q) = 1 + tq \quad \text{if } q = \langle\alpha\rangle \text{ with } \alpha \in k^*,$$

characterize $\lambda_t(q)$.

(2) Let $n \in \mathbf{Z}$ be the image of q under the rank map $WGr(k) \to \mathbf{Z}$. Use (1) to prove that

$$\frac{\mathrm{d}(\lambda_t(q))/\mathrm{d}t}{\lambda_t(q)} = \frac{q - nt^2}{1-t^2}.$$

Deduce from this formula that the powers q^i of q can be written as linear combinations (with integral coefficients depending on n) of the $\lambda^{i-2d}(q)$, e.g.:

$$q^2 = 2\lambda^2(q) + n$$
$$q^3 = 6\lambda^3(q) + (3n-2)q$$

(3) If u, v are indeterminates, show that

$$\lambda_u(q)\cdot\lambda_v(q) = (1+uv)^n\cdot\lambda_{(u+v)/(1+uv)}(q)$$

in the ring $WGr(k)[[u,v]]$. Use this identity to express the products $\lambda^i(q)\cdot\lambda^j(q)$ as $\mathbf{Z}$-linear combinations of the $\lambda^{i+j-2a}(q)$, $0 \le a \le \sup(i,j)$.

27.3. WITT INVARIANTS. A morphism of functors $\mathrm{Torsors}_G \to W$ is called a *Witt invariant* (of G-torsors). Almost all of what we have done for cohomology and cohomological invariants works for W and Witt invariants. For example:

27.4. LOCAL CASE. Let K be complete with respect to a discrete valuation v with residue field k (of char $\neq 2$), and let π be a uniformizing element of K.

There is a natural embedding $W(k) \to W(K)$, which maps $\langle a_1,\dots,a_n\rangle$ (or rather its class), with $a_i \in k^*$, to $\langle u_1,\dots,u_n\rangle$, where the u_i are units of K lifting the a_i. If one identifies the ring $W(k)$ with its image in $W(K)$, then one has a direct sum decomposition:

THEOREM 27.5. (Springer, see e.g. [Lam 73, VI.1.5]) *The ring $W(K)$ is a free $W(k)$-module of rank 2, with basis $\{1, \langle\pi\rangle\}$. In particular, one has*

$$W(K) = W(k) \oplus \langle\pi\rangle\cdot W(k).$$

If $q = q_1 + \langle\pi\rangle\cdot q_2$ is an element of $W(K)$ with q_1, $q_2 \in W(k)$, then q_1 is called the "first residue" of q and q_2 the "second residue" of q. In what follows, the second residue will be called simply the *residue*; it will be denoted by $\partial_\pi(q)$. We have an exact sequence

$$0 \to W(k) \to W(K) \xrightarrow{\partial_\pi} W(k) \to 0.$$

(Although the residue map ∂_π depends on the choice of π, this choice has no influence on the applications we give later.[a])

An element of $W(K)$ is called *unramified* if its (second) residue is 0, i.e., if it belongs to the subring $W(k)$ of $W(K)$.

27.6. THE NON-COMPLETE CASE. The definitions of 27.4 are extended to the non-complete case as follows:

---

[a]Note the similarity and the differences with the cohomological residue of §7. In the cohomology situation, we got an intrinsic residue, but at the cost of a Tate twisting on the module. Here, we have a residue depending on the choice of π; it could also be interpreted as a canonical residue with values in a suitable twist of $W(k)$.

Let K be a field with a discrete valuation v, ring of integers R, uniformizing element π, and residue field $k(v) = R/\pi R$; let K_v be its completion. If $q \in W(K)$, its residue $\partial_\pi(q) \in W(k(v))$ is the image of q under the map

$$W(K) \to W(K_v) \xrightarrow{\partial_\pi} W(k(v)).$$

If it is 0, q is said to be *unramified at* v; its image in $W(K_v)$ belongs to $W(k(v))$ and is called the *value* of q at v.

27.7. WITT RING OF $K = k(t)$. With the same notation as in Section 9, we have an exact sequence (due to Tate and Milnor, cf. [Mi 70, Th. 2.3])

$$0 \to W(k) \to W(K) \xrightarrow{\partial} \oplus_{v \neq \infty} W(k(v)) \to 0$$

where ∂ is given by the residues (the choice of the uniformizing elements does not matter).

In particular, an element of $W(K)$ is *constant* (i.e., belongs to $W(k)$) if and only if its residues at all $v \neq \infty$ are 0.

Moreover, an element q of $W(K)$ which is *unramified outside 0 and* ∞ (i.e., with zero residues for $v \neq 0, \infty$) can be written uniquely as $q = q_0 + q_1\langle t\rangle$, with q_0, $q_1 \in W(k)$; of course, q_1 is the residue of q at 0 with respect to the choice of t as uniformizing element.

27.8. $K = k(t_1, \ldots, t_n)$. By induction on n and starting with 27.7, one sees that if $q \in W(K)$ has zero residues for all the discrete valuations of K coming from irreducible hypersurfaces of $\mathrm{Aff}^n_{/k}$, then q is *constant.*

27.9. COMPATIBILITY AND SPECIALIZATION THEOREMS. All the statements of Chapter IV extend with essentially the same proofs. More precisely, for 27.10 through 27.13, let G be a smooth algebraic group over k_0, and let a be an element of $\mathrm{Inv}(G, W) = \mathrm{Inv}(\mathrm{Torsors}_G, W)$. Then:

THEOREM 27.10. *If K is local as in §11, the diagram*

$$\begin{array}{ccc} H^1(k,G) & \longrightarrow & H^1(K,G) \\ \downarrow a & & \downarrow a \\ W(k) & \longrightarrow & W(K) \end{array}$$

is commutative.

PROOF. Same as for Theorem 11.1. □

THEOREM 27.11. *Let (K, v) be as in 27.6 above (K not necessarily complete), with $k_0 \subset R$, and let T be an R-G-torsor. Write T_K and $T_{k(v)}$ for the K-G-torsor and the $k(v)$-G-torsor deduced from T by the base changes $R \to K$ and $R \to k(v)$. Put $\alpha = a(T_K)$ in $W(K)$. Then:*

(1) *α is unramified at v.*
(2) *The value of α at v is $a(T_k)$.*

This follows from 27.10 by replacing K with its completion K_v at v.

THEOREM 27.12. *Let R_{m}, K, k, and T be as in* 12.1 *and* 12.2. *Then* $a(T_K) = 0$ *implies* $a(T_k) = 0$.

PROOF. Same as for Th. 12.2: by induction on $\dim R_{\mathrm{m}}$, starting from the case $\dim R_{\mathrm{m}} = 1$, which is 27.11. □

COROLLARY 27.13. *If T is versal, then $a(T) = 0 \Rightarrow a = 0$.*

27.14. WITT INVARIANTS OF $(2, \ldots, 2)$ GROUPS.
Let $G = \mathbf{Z}/2\mathbf{Z} \times \cdots \times \mathbf{Z}/2\mathbf{Z}$ (n copies). An arbitrary element of $H^1(k, G)$ is given by an n-tuple $(\alpha_1, \ldots, \alpha_n) \in k^*/k^{*2} \times \cdots \times k^*/k^{*2}$. For I a subset of $[1, n]$, we write $\langle \alpha_I \rangle$ where α_I is the product of α_i for $i \in I$ as in 27.1.

Write a_I for the Witt invariant $(\alpha_1, \ldots, \alpha_n) \mapsto \langle \alpha_I \rangle$.

THEOREM 27.15. *$\mathrm{Inv}_{k_0}(G, W)$ is a free $W(k_0)$-module with basis $(a_I)_{I \subset [1,n]}$.*

PROOF. Suppose first that $n = 1$, i.e., $G = \mathbf{Z}/2\mathbf{Z}$. Let $k = k_0(t)$, and let T be the k-G-torsor defined by t, as in the proof of 16.2. If $a \in \mathrm{Inv}_{k_0}(G, W)$, then $a(T)$ is an element of $W(k)$ which is unramified outside $\{0, \infty\}$ by 27.11. By 27.7, $a(T)$ is of the form $q_0 \oplus q_1 \cdot \langle t \rangle$, with q_0, $q_1 \in W(k_0)$. By 27.13 this implies that $a = q_0 \oplus q_1 \cdot a_1$, where a_1 is the tautological invariant $(\alpha) \mapsto \langle \alpha \rangle$. This proves the theorem for $n = 1$.

The general case follows by induction on n, as in the proof of Th. 16.4. □

THEOREM 27.16 (Witt invariants of Quad_n). *$\mathrm{Inv}(\mathrm{Quad}_n, W)$ is a free $W(k_0)$-module with basis the exterior powers $\{\lambda^0, \lambda^1, \ldots, \lambda^n\}$.*

In other words, *any invariant of a rank n quadratic form q can be written uniquely as*

$$q \mapsto \sum_{i=0}^{n} c_i \lambda^i(q), \quad \text{with } c_i \in W(k_0).$$

PROOF OF TH. 27.16. Essentially the same as for Th. 17.3: The standard embedding $G \to O_n$ gives a restriction map (cf. §13)

$$\mathrm{Inv}(\mathrm{Quad}_n, W) \to \mathrm{Inv}(G, W)$$

which is injective. By Prop. 13.2, its image is contained in the $W(k_0)$-submodule of $\mathrm{Inv}(G, W)$ fixed by the action of S_n; this submodule has a basis made up of the $\lambda^i = \sum_{|I|=i} a_I$ for $i = 0, 1, \ldots, n$. The result follows. □

EXAMPLE 27.17 (Witt invariants of $\mathrm{Pfister}_n$ and Oct). $\mathrm{Inv}(\mathrm{Pfister}_n, W)$ is a free $W(k_0)$-module with basis 1 and the identity. The case $n = 3$ gives that $\mathrm{Inv}(\mathrm{Oct}, W)$ is a free $W(k_0)$-module with basis $\{1, q_3\}$, where q_3 is the norm of the octonion algebra.

EXAMPLE 27.18 (Witt invariants of Alb). Similarly, $\mathrm{Inv}(\mathrm{Alb}, W)$ is a free $W(k_0)$-module with basis $\{1, q_3, q_5\}$, where $q_3(A)$, $q_5(A)$ are the 3-Pfister and 5-Pfister forms associated with the Jordan algebra A as in 22.4.

EXERCISE 27.19. Let n be even and ≥ 4; set $m = n/2$. Let q be a quadratic form of rank n, and pick $a \in k^*$ represented by q. We may write q as $q = \langle a \rangle \oplus q_a$, with $\mathrm{rank}(q_a) = n - 1$. Put:

$$\ell(a, q) = \lambda^m(q_a)$$

Then:

(1) If $d(q) = 1$, prove that $\ell(a, q)$ does not depend on the choice of the element a represented by q. Hence $\ell(a, q)$ is an invariant of q, which we denote by $\ell(q)$. One has $2\ell(q) = \lambda^m(q)$, so that $\ell(q)$ may be seen as a kind of "canonical half" of $\lambda^m(q)$.

(2) Do not assume that $d(q) = 1$, but choose an element $r \in W(k_0)$ such that $r\cdot\langle 1, -d(q)\rangle = 0$ in $W(k_0)$. Show that the product $r\cdot\ell(a,q)$ is independent of the choice of a; it is a Witt invariant of $\mathrm{Quad}_{n,d(q)}$, which we may denote by $\ell_r(q)$.

(3) With the notation of (2), show that every Witt invariant of $\mathrm{Quad}_{n,\delta}$ can be written in a unique way as

$$q \mapsto \ell_r(q) + \sum_{i=0}^{m-1} r_i\cdot\lambda^i(q),$$

where the r_i and r belong to $W(k_0)$ and the product $r\cdot\langle 1, -\delta\rangle$ is 0 in $W(k_0)$.

[Hints: (1) may be proved by the usual chain device. The same applies to (2), or (2) may be deduced from (1). The proof of (3) is similar to that of Th. 20.6.]

EXAMPLE 27.20. The Lie algebra of derivations of an octonion (resp. Albert) algebra is of type G_2 (resp. F_4). If the characteristic is $\neq 2, 3$, its Killing form is nondegenerate and gives an invariant $\mathrm{Oct} \to W$ (resp. $\mathrm{Alb} \to W$). By 27.17, this invariant can be written in terms of the invariant q_3 (resp. the invariants q_3 and q_5). When one does the computation, one finds:

G_2 CASE: $\text{Killing} = \langle -1, -3\rangle(q_3 - 1)$

F_4 CASE: $\text{Killing} = \langle -2\rangle(q_5 - q_3) + \langle -1, -1, -1, -1\rangle(q_3 - 1)$

(Here, and in what follows, if q is a quadratic form of rank n which represents 1, we write $q - 1$ for the quadratic form q' of rank $n - 1$ such that $q = \langle 1\rangle \oplus q'$. The quadratic form $q_5 - q_3$ is defined similarly, cf. Th. 22.4.)

The formula for F_4 can be found in [Lo 01, (33)] or can be deduced from [Ja 71, p. 111, (133)].

EXERCISE 27.21. Keep the notation above, but assume the characteristic is 3.

(1) In the G_2 case, prove that the Killing form has rank 7 instead of 14 and that it is isomorphic to $\langle -1\rangle(q_3 - 1)$.

(2) In the F_4 case, prove that the Killing form is 0, but that there exists a nondegenerate invariant quadratic form on the Lie algebra which is isomorphic to $(q_5 - 1) \oplus \langle 1, 1, 1\rangle(q_3 - 1)$. [Hint: use the "normalized Killing form", i.e., lift to the Witt vectors, divide the Killing form by 36, and reduce mod 3.]

27.22. WITT INVARIANTS OF Herm_n. Recall that we consider hermitian forms relative to a quadratic extension $k_1 = k_0(\sqrt{\delta})$ of k_0. Fix an $i \geq 1$ and an element $x \in W(k_0)$ such that $x \in \langle 1, -\delta\rangle W(k_0) = \ker[W(k_0) \to W(k_1)]$. If $h \in \mathrm{Herm}_n(k)$ is represented by $q \in \mathrm{Quad}_n(k)$, we have $x\cdot\lambda^i(q) \in W(k)$. Moreover:

LEMMA. *The product $x\cdot\lambda^i(q)$ depends only on x and h.*

PROOF. Write x as $\langle 1, -\delta\rangle y$, and observe that

$$x\cdot\lambda^i(q) = y\cdot\langle 1, -\delta\rangle\cdot\lambda^i(q) = y\cdot N\lambda^i h,$$

where now $\lambda^i h$ is the i-th exterior product of h (in the sense of hermitian forms) and $N\colon \mathrm{Herm}_n \to \mathrm{Quad}_{2n}$ is as defined in 21.1. □

We may then define $\lambda^i_x(h)$ as being $x\cdot\lambda^i(h)$.

THEOREM. *Every Witt invariant of* Herm_n *can be written uniquely as*

$$h \mapsto x_0 + \sum_{i=1}^{n} \lambda^i_{x_i}(h), \quad \textit{with} \begin{cases} x_0 \in W(k_0) \\ x_i \in \langle 1, -\delta\rangle\cdot W(k_0) \quad \textit{if } i \geq 1. \end{cases}$$

The proof is the same as that of Theorem 21.6.

REMARK 27.23. The theorem above, as well as Theorem 21.6 giving the cohomological invariants of Herm_n, can be stated as

$$\mathrm{Inv}^{\mathrm{norm}}\, \mathrm{Herm}_{n,k_1/k_0} = \ker\left[\mathrm{Inv}^{\mathrm{norm}}\, \mathrm{Quad}_{n/k_0} \to \mathrm{Inv}^{\mathrm{norm}}\, \mathrm{Quad}_{n/k_1}\right],$$

where $\mathrm{Inv}^{\mathrm{norm}}$ denotes the normalized invariants.

28. Quadratic form invariants of $\mathrm{Pfister}_n$ and Quad_n

In this section, we consider morphisms $A \to \mathrm{Quad}_N$, which we call *quadratic form invariants*; we write $\mathrm{Inv}(A, \mathrm{Quad}_N)$ for the set of all such morphisms. (Note that these are not strictly invariants in the sense of §1 because Quad_N takes values in the category Sets instead of the category Abelian Groups.) First an auxiliary result:

PROPOSITION 28.1. *Let $x_1, \ldots, x_n$ be indeterminates, and $k = k_0(x_1, \ldots, x_n)$. For each $I \subset [1, n]$ let x_I denote $\prod_{i\in I} x_i$, and let φ_I be an anisotropic quadratic form over k_0. Then the quadratic form*

$$\sum_I \varphi_I \cdot \langle x_I \rangle$$

is anisotropic over k.

PROOF. The case $n = 1$ is standard. The general case follows by induction. □

COROLLARY 28.2. *Let φ_0 and φ_1 be anisotropic quadratic forms over k_0. For each $n \geq 1$, there is an n-Pfister form q defined over an extension k of k_0 with $\varphi_0 + \varphi_1 \cdot (q - 1)$ anisotropic over k.*

Recall from 27.20 that, if q is a quadratic form of rank n which represents 1, we write $q - 1$ for the quadratic form q' of rank $n - 1$ such that $q = \langle 1 \rangle \oplus q'$.

PROOF. Set $k = k_0(x_1, \ldots, x_n)$ for $x_1, \ldots, x_n$ indeterminates, and set $q = \langle 1, x_1 \rangle \otimes \cdots \otimes \langle 1, x_n \rangle$. In the notation of Proposition 28.1,

$$\varphi_0 + \varphi_1 \cdot (q - 1) = \sum_I \varphi_I \cdot \langle x_I \rangle, \text{ with } \varphi_I = \begin{cases} \varphi_0 & \text{if } I = \emptyset \\ \varphi_1 & \text{if } I \neq \emptyset. \end{cases}$$

Proposition 28.1 gives the result. □

COROLLARY 28.3. *Let φ_0, …, φ_n be anisotropic quadratic forms over k_0. There exists an n-dimensional quadratic form q defined over an extension k of k_0 with $\sum_{i=0}^n \varphi_i \cdot \lambda^i(q)$ anisotropic over k.*

PROOF. Set $k = k_0(x_1, \ldots, x_n)$ for $x_1, \ldots, x_n$ indeterminates, and set $q = \langle x_1, \ldots, x_n \rangle$. The claim is now reduced to Proposition 28.1, just as the proof of Corollary 28.2 did. □

THEOREM 28.4. *Every invariant in* $\mathrm{Inv}(\mathrm{Pfister}_n, \mathrm{Quad}_N)$ *can be written uniquely as*

$$q \mapsto \varphi_0 \oplus \varphi_1 \cdot (q - 1)$$

with φ_0, $\varphi_1 \in \mathrm{Quad}(k_0)$, with φ_1 anisotropic.

PROOF. Let a be an invariant $\text{Pfister}_n \to \text{Quad}_N$. Composing it with the map $\text{Quad}_N \to W$ gives an invariant b in $\text{Inv}(\text{Pfister}_n, W)$. By 27.17, b may be written as

$$b(q) = f_0 + f_1 \cdot (q - 1) \quad \text{in } W(k)$$

for some forms f_0, $f_1 \in \text{Quad}(k_0)$, which one may choose to be anisotropic. Put $N' = \text{rank}(f_0) + \text{rank}(f_1) \cdot (2^n - 1)$. Since $a(q)$ and $b(q)$ have the same rank (mod 2), we have $N' \equiv N \pmod 2$. There are two cases:

Case I: $N' \leq N$. Put $h = (N - N')/2$. Define quadratic forms φ_0 and φ_1 by:

$$\varphi_0 = f_0 \oplus h \cdot \langle 1, -1 \rangle, \qquad \varphi_1 = f_1.$$

We have $\text{rank}(\varphi_0 \oplus \varphi_1 \cdot (q-1)) = N' + 2h = N$. Since $a(q)$ and $\varphi_0 \oplus \varphi_1 \cdot (q - 1)$ have the same rank and the same image in $W(k)$, they are equal. This gives the formula we wanted for $a(q)$. Its uniqueness is clear.

Case II: $N' > N$. Put $h = (N' - N)/2$. The same argument as above shows that, for every k/k_0, and every $q \in \text{Pfister}_n(k)$, the quadratic form $f_0 \oplus f_1 \cdot (q - 1)$ contains $h \cdot \langle 1, -1 \rangle$, hence is isotropic. This contradicts Cor. 28.2, applied to f_0 and f_1. Hence this case is impossible. □

THEOREM 28.5. *Every invariant in* $\text{Inv}(\text{Quad}_n, \text{Quad}_N)$ *can be written uniquely as*

$$q \mapsto \varphi_0 + \varphi_1 \cdot q + \cdots + \varphi_m \cdot \lambda^m q,$$

where $\varphi_i \in \text{Quad}(k_0)$ *and* $\varphi_1, \ldots, \varphi_m$ *are anisotropic.*

(Note that $\varphi_1, \ldots, \varphi_m$ are allowed to be 0-dimensional.)

PROOF. The proof is analogous to the proof of Theorem 28.4, except that one uses Cor. 28.3 instead of Cor. 28.2. □

29. Witt invariants of étale algebras

THEOREM 29.1. *A Witt invariant for* Et_n *is 0 if it is 0 for all multiquadratic algebras.*

PROOF. We use induction on n. Let $a \in \text{Inv}(\text{Et}_n, W)$ be such that $a(E) = 0$ for all multiquadratic algebras E. Using the induction assumption one shows, as in 24.9, that a vanishes on every étale algebra which has a factor of rank 1 or 2. Next one observes that the invariant $a(E^{\text{gen}})$ of the versal torsor E^{gen} of 24.6 is unramified (i.e., has residue 0) at all the irreducible divisors of Aff^n, except perhaps Δ. This follows from 27.12. Furthermore, the versal torsor splits into the product of a quadratic algebra and something else "locally at Δ" (i.e., over the local field K_Δ, as in 24.6) The induction assumption then shows that $a(E^{\text{gen}})$ gives 0 in $W(K_\Delta)$, hence has residue 0 at Δ as well.

By 27.8, $a(E^{\text{gen}})$ belongs to $W(k_0)$. Since E^{gen} is versal, a is constant by 27.13. Since a is 0 on multiquadratic algebras, it is identically 0. □

THEOREM 29.2. *Set* $m = [n/2]$. *Then* $\text{Inv}(\text{Et}_n, W)$ *is a free* $W(k_0)$*-module with basis the invariants*

$$E \mapsto \lambda^i(q_E) \text{ for } 0 \leq i \leq m,$$

where q_E *is the trace form of* E.

PROOF. Assume that n is even, so that $n = 2m$. (The case where n is odd is similar.) Let a be an element of $\mathrm{Inv}(\mathrm{Et}_n, W)$. As in 24.10 and 25.5, let H be the product of m copies of $\mathbf{Z}/2\mathbf{Z}$, viewed as the subgroup of S_n generated by the transpositions (1 2), (3 4), ... An element α of $H^1(k, H)$ is (cf. 27.14) an m-tuple $(\alpha_1, \dots, \alpha_m)$ and defines a multiquadratic algebra $E_\alpha = \prod_i E_{\alpha_i}$ for E_{α_i} the quadratic algebra $k[X]/(X^2 - \alpha_i)$. The map $\alpha \mapsto a(E_\alpha)$ is a Witt invariant of $H^1(*, H)$, hence (by Th. 27.14) can be written uniquely as a linear combination

$$\sum c_I \cdot a_I(\alpha) = \sum c_I \cdot \langle \alpha_I \rangle \quad \text{with } c_I \text{ in } W(k_0).$$

By using the action of S_m on G, one sees (as in the proof of Th. 17.3) that c_I depends only on the number of elements d of I, so that we may rewrite the above sum as

$$\sum_{d=0}^{m} c_d \cdot b_d(\alpha) \quad \text{with } b_d(\alpha) = \sum_{|I|=d} \langle \alpha_I \rangle.$$

On the other hand, the trace form q of the algebra E_α is given by

$$\langle 2 \rangle \cdot q = \langle 1, \dots, 1 \rangle \oplus \langle \alpha_1, \dots, \alpha_m \rangle.$$

One deduces from this that the exterior powers $\lambda^d(q)$ for $d = 0, \dots, m$ are related to the $b_d(\alpha)$ by a triangular invertible matrix with coefficients in $W(k_0)$, where these coefficients depend only on m. For example,

$$\lambda^0(q) = b_0(\alpha)$$
$$\lambda^1(q) = \langle 2 \rangle \cdot b_1(\alpha) + m \cdot \langle 2 \rangle \cdot b_0(\alpha)$$
$$\lambda^2(q) = b_2(\alpha) + m \cdot \langle 2 \rangle \cdot b_1(\alpha) + \frac{m(m-1)}{2} \cdot b_0(\alpha)$$

This shows that there exists a unique choice of elements $c'_d \in W(k_0)$ such that

$$a(E) = \sum c'_d \cdot \lambda^d(q_E)$$

when E is multiquadratic, hence for every étale algebra E by Th. 29.1. □

Note that in 29.2, i runs through the integers $\leq m = [n/2]$. This implies in particular that the $\lambda^i(q_E)$, $i > m$, can be written as linear combinations of 1, $\lambda^1(q_E)$, ..., $\lambda^m(q_E)$. This can also be deduced from the following more precise fact. Define a quadratic form ω_n by

(29.3) $$\omega_n = \begin{cases} \langle 2, 2, \dots, 2 \rangle \text{ (rank } m) & \text{if } n = 2m \\ \langle 1, 2, \dots, 2 \rangle \text{ (rank } m+1) & \text{if } n = 2m+1. \end{cases}$$

THEOREM 29.4. *$\lambda^i(q_E - \omega_n) = 0$ for all $i > m$.*

(Here $q_E - \omega_n$ is viewed as an element of $WGr(k)$, so that $\lambda^i(q_E - \omega_n)$ makes sense.)

PROOF. By Theorem 29.1, it is sufficient to prove this when the étale algebra E is multiquadratic. But in that case q_E is of the form

$$q_E = \omega_n \oplus Q,$$

where Q is a quadratic form of rank m. It is then obvious that $\lambda^i Q = 0$ for all $i > m$, and the result follows. □

Theorem 29.4 gives:

$$\sum_{a+b=i} \lambda^a(q_E)\cdot\lambda^b(-\omega_n) = 0 \quad \text{for } i > m.$$

This gives a recursive process for computing the higher exterior powers of q_E in terms of the lower ones, see e.g. (30.3) and (30.4) in the cases $n = 4$ and $i = 3, 4$.

REMARK 29.5. Define $S_E \in WGr(k)$ as $q_E - \omega_n$; one has $\operatorname{rank}(S_E) = m$. It is not always true (cf. 31.8) that S_E is an "effective" element of $WGr(k)$, i.e., that it can be represented by a genuine quadratic form of rank m. Still, Th. 29.4 shows that the λ-powers of S_E behave as if S_E were effective; this has been recently completed by Rost [Rost 02], who has shown that there exists an m-Pfister form $q_m(E)$ such that

$$q_m(E) = \oplus_{i=0}^{m} \lambda^i(q_E - \omega_n) \quad \text{in } WGr(k).$$

(Note that the existence of $q_m(E)$ would be obvious if S_E were effective.) This gives a Pfister form invariant of étale algebras.

30. Applications

The fact that multiquadratic algebras detect the equality of Witt invariants can be used to perform explicit computations. For instance, let $\varphi\colon S_n \to S_N$ be a homomorphism of symmetric groups. This gives a morphism of functors $\mathrm{Et}_n \to \mathrm{Et}_N$, and we write E_φ for the image of E, which is well-defined up to isomorphism.[b] In classical language, E_φ corresponds to a "resolvent" of E. For instance, the obvious map $S_4 \to S_3$ allows one to attach to a quartic algebra E its "cubic resolvent" E_3.

One may ask the following question: *Is there a formula giving the trace form q_{E_φ} in terms of q_E?* The answer is yes: Indeed, the map $E \mapsto q_{E_\varphi}$ is a Witt invariant of E over k_0 = the prime field. Hence by Theorem 29.2 one can write q_{E_φ} *as a linear combination of the* $\lambda^i q_E$ with coefficients in $W(k_0)$. And these coefficients can be computed by looking at what happens when E is multiquadratic. Here are three examples.

30.1. THE CUBIC RESOLVENT OF A QUARTIC EQUATION. Let E be an étale algebra of rank 4 and let E' be the algebra of rank 3 deduced from it via the map $S_4 \to S_3$ mentioned above. One wants to compute the trace form $q_{E'}$ in terms of the trace form $q = q_E$. Any trace form of rank 3 is equal to $\langle 1, 2, 2d\rangle$ where d is the discriminant, see Theorem 31.10. Since $S_4 \to S_3$ is compatible with the sign homomorphisms $S_n \to \mathbf{Z}/2\mathbf{Z}$, d is also the discriminant of E. Hence we get $q_{E'} = \langle 1, 2, 2d\rangle$ with $d = d(q)$. Note that we could write $\langle d\rangle$ as $\lambda^4 q$, and thus:

(30.2) $$q_{E'} = \langle 1, 2\rangle \oplus \langle 2\rangle\lambda^4 q.$$

However, this is not written as a linear combination of 1, q, and $\lambda^2 q$, as it should. To do so, one has to use the identities of Theorem 29.4, which tell us that $\lambda^i(q-2) = 0$ for $i = 3, 4$ (since $\omega_2 = \langle 2, 2\rangle = \langle 1, 1\rangle = 2$). That is,

(30.3) $$\lambda^3 q - 2\lambda^2 q + 3q - 4 = 0$$

(30.4) $$\lambda^4 q - 2\lambda^3 q + 3\lambda^2 q - 4q + 5 = 0.$$

[b]More generally, if X is a finite étale covering of rank n of a scheme S, one can associate with it — in a functorial way — a finite étale scheme X_ϕ of rank N. [Indeed, X defines an S_n-torsor P_X over S, and X_ϕ is the fiber space associated with P_X and the ϕ-action of S_n on the finite set $\{1, ..., N\}$.]

By combining (30.2), (30.3), and (30.4), one obtains the required formula:

PROPOSITION 30.5. $q_{E'} = 5 - 2q_E + \langle 2\rangle\lambda^2 q_E$

Note that the formula holds in $WGr(k)$, not just in $W(k)$, since both sides have rank 3.

REMARK 30.6. If k contains the 8th roots of unity, we have $\langle 2\rangle = \langle -1\rangle = 1$ and $2 = 0$ in $W(k)$, so that 30.5 can be rewritten in a simplified way:

$$q_{E'} = 1 + \lambda^2 q_E \quad \text{in } W(k). \tag{30.5$'$}$$

30.7. THE SEXTIC RESOLVENT OF A SEXTIC EQUATION. Let E be an étale algebra of rank 6, and let E' be the étale algebra obtained from E via an outer automorphism of S_6 ("sextic resolvent"). Let q and q' denote the respective trace forms.

By Theorem 29.1, in order to describe q' in terms of q, it is sufficient to do so when E is multiquadratic, i.e., when E splits as a product of three quadratic algebras:

$$E = E_a \times E_b \times E_c,$$

where $E_a = k[x]/(x^2 - a)$, ..., $E_c = k[x]/(x^2 - c)$ with a, b, $c \in k^*$. Then

$$q = \langle 2\rangle\langle 1,1,1,a,b,c\rangle = 3\langle 2\rangle \oplus \langle 2\rangle\langle a,b,c\rangle. \tag{30.8}$$

Let $C = \mathbf{Z}/2\mathbf{Z} \times \mathbf{Z}/2\mathbf{Z} \times \mathbf{Z}/2\mathbf{Z}$ and identify C with its image in S_6 under the embedding corresponding to the partition of $\{1,\dots,6\}$ as $\{1,2\} \cup \{3,4\} \cup \{5,6\}$. To determine the structure of E', we have to look at the image C' of C under an outer automorphism of S_6.

Such an automorphism transforms a transposition into a product of three disjoint transpositions and conversely. By construction, the group C contains only one product of three disjoint transpositions, represented by the diagonal subgroup D. Thus the group C' contains only one transposition, which generates the image D' of D. This gives a decomposition of $\{1,\dots,6\}$ into two orbits (say) $\{1,2\}$ and $\{3,\dots,6\}$, where the action on $\{3,\dots,6\}$ is the free action of $C'/D' \cong \mathbf{Z}/2\mathbf{Z}\times\mathbf{Z}/2\mathbf{Z}$. This corresponds to a decomposition of E' as a direct sum of the quadratic algebra E_{abc} and the biquadratic Galois algebra $E_{ab}\otimes E_{ac}$ (which is isomorphic to $E_{ab}\otimes E_{bc}$ and $E_{ac}\otimes E_{bc}$). That is,

$$E' = E_{abc} \times (E_{ab} \otimes E_{ac}).$$

Hence:

$$q' = \langle 2, 2abc\rangle \oplus \langle 2, 2ab\rangle \otimes \langle 2, 2ac\rangle = \langle 2, 2abc\rangle \oplus \langle 1, ab, ac, bc\rangle.$$

Setting

$$\gamma_1 = \langle a,b,c\rangle, \quad \gamma_2 = \langle ab,ac,bc\rangle, \quad \gamma_3 = \langle abc\rangle,$$

one has:

$$\begin{aligned} q &= \langle 2\rangle(3+\gamma_1) \\ \lambda^2 q &= 3 + 3\gamma_1 + \gamma_2 \\ \lambda^3 q &= \langle 2\rangle(1 + 3\gamma_1 + 3\gamma_2 + \gamma_3) \\ q' &= \langle 1,2\rangle + \gamma_2 + \langle 2\rangle\gamma_3. \end{aligned}$$

This implies:

PROPOSITION 30.9. $q' = -\langle 1,2\rangle + \langle 1,1,2\rangle q - \langle 1,2\rangle\lambda^2 q + \lambda^3 q$

REMARKS 30.10. (1) As in 30.6, if we assume that k contains the 8th roots of unity, 30.9 simplifies as:

$$q' = q + \lambda^3 q \quad \text{in } W(k). \tag{30.9$'$}$$

(2) By Theorem 31.26, q can be written as $\langle 1,2\rangle \oplus Q$ where Q is a quadratic form of rank 4. Similarly, q' can be written as $\langle 1,2\rangle \oplus Q'$. Hence 30.9 can be rewritten in terms of Q and Q'. The result is much simpler, namely:

$$Q' = \lambda^3 Q = \langle 2d\rangle Q, \text{ where } d = d(q) \text{ is the discriminant of } q. \tag{30.11}$$

30.12. ANOTHER EXAMPLE. Consider the subsets of $\{1,\ldots,n\}$ with two elements. The action of S_n on these subsets gives a homomorphism

$$\varphi\colon S_n \to S_N, \quad \text{where } N = \binom{n}{2}.$$

(This homomorphism is only defined up to conjugation.) As above, this gives a map $\mathrm{Et}_n \to \mathrm{Et}_N$; call $E(2)$ the étale algebra of rank N corresponding to an E in $\mathrm{Et}_n(k)$. (In classical terms, this corresponds to taking the sums $x_i + x_j$ of the roots of an equation, provided the equation is general enough. Scheme-theoretically, if X is a finite étale covering of S, one defines $X(2)$ by removing the diagonal from $X \times_S X$ and dividing by the involution $(x,y) \mapsto (y,x)$.) By §29, one knows that the trace form $q_{E(2)}$ can be computed in terms of the exterior powers $\lambda^i(q_E)$. More precisely:

PROPOSITION 30.13. *One has the formula:*

$$q_{E(2)} = \lambda^2(q_E) - \langle 2\rangle(q_E - n).$$

PROOF. The formula holds when $\operatorname{rank} E = 1$ (where $E(2) = 0$).

When $\operatorname{rank} E = 2$, we have $E(2) = k$, $q_{E(2)} = 1$, and $q_E = \langle 2, 2\alpha\rangle$ for some $\alpha \in k^*$. Then $\lambda^2(q_E) = \langle\alpha\rangle$ and

$$\langle 2\rangle(q_E - n) = \langle 1,\alpha\rangle - \langle 2,2\rangle = \langle\alpha\rangle - 1,$$

since $\langle 2,2\rangle = \langle 1,1\rangle$. Hence the formula holds in this case.

For $E = E' \times E''$, one has the formula

$$E(2) \cong E'(2) \times (E' \otimes E'') \times E''(2),$$

which implies a corresponding equality for trace forms:

$$q_{E(2)} = q_{E'(2)} \oplus q_E \cdot q_{E'} \oplus q_{E''(2)}.$$

This shows by induction on n that the proposition holds when E is multiquadratic. By Th. 29.1, it holds in general. □

REMARKS 30.14. (1) The formula for the trace form of $E(3)$ (defined in an obvious way) is

$$q_{E(3)} = \lambda^3(q) - 2\cdot\lambda^2(q) + n\cdot\langle 2\rangle\cdot(q-1),$$

where $q = q_E$ and $n = \operatorname{rank} E$.

(2) A different proof of Proposition 30.13 was suggested by M. Knus: Write $(E \otimes E)^+$ for the subalgebra of $E \otimes E$ consisting of elements fixed by the involution which permutes the factors (i.e., $(E \otimes E)^+$ is $\mathsf{TS}^2(E)$, where TS means "symmetric tensors", not to be confused with the symmetric powers, which are quotients). This is an étale algebra. The obvious homomorphism $E \otimes E \to E$ restricts to a surjective map $(E \otimes E)^+ \to E$, hence E is a direct summand of $(E \otimes E)^+$. Indeed, $(E \otimes E)^+$ is isomorphic to $E \times E(2)$. This decomposition is compatible with trace forms, hence

$$\mathsf{TS}^2(q) = q + \langle 2 \rangle \cdot q_{E(2)}$$

where $\mathsf{TS}^2(q)$ is the symmetric square of q, i.e., the restriction of $q \otimes q$ to the symmetric part of $E \otimes E$. By taking orthogonal bases, one sees that

$$\mathsf{TS}^2(q) = n + \langle 2 \rangle \cdot \lambda^2(q).$$

This gives the desired formula

$$q_{E(2)} = \lambda^2(q) - \langle 2 \rangle (q - n).$$

CHAPTER IX

The trace form in dimension ≤ 7

In this chapter (except in 33.1 through 33.13) the characteristic is assumed to be $\neq 2$.

As in Chapter VI, we put $H^i(k) = H^i(k, \mathbf{Z}/2\mathbf{Z})$ and $H(k) = \oplus H^i(k)$.

31. Characterization of trace forms in dimension ≤ 7

In this section, q is a quadratic form over k of rank n. For $n \leq 7$, we give a criterion for q to be isomorphic to a trace form q_E where E is an étale algebra of rank n over k.

31.1. A NECESSARY CONDITION. (Cf. [Se 84, App. I]) Let E be an étale algebra over k of rank n. Write n in dyadic form as $n = \sum_{i=1}^{h} 2^{m_i}$ for $0 \leq m_1 < \cdots < m_h$.

PROPOSITION 31.2. *There exists a finite extension k'/k of odd degree such that $E' = k' \otimes E$ splits as a product of étale algebras E'_i with* rank $E'_i = 2^{m_i}$.

PROOF. By using Sylow's theorem for the prime 2, one shows that k'/k can be chosen such that the "Galois group" of $E' = k' \otimes E$ (i.e., the image of $\phi_{E'}$) is a 2-group. The result follows by decomposing $[1, n]$ into the orbits of this group, cf. [Se 84, Lemma 3]. □

Let us now define a quadratic form r_n of rank h (where h is as above) by the formula

$$r_n = \langle 2^{m_1}, \ldots, 2^{m_h} \rangle. \tag{31.3}$$

Note that $\langle 2^m \rangle$ is isomorphic to $\langle 1 \rangle$ if m is even and to $\langle 2 \rangle$ if m is odd; moreover $\langle 2, 2 \rangle$ is isomorphic to $\langle 1, 1 \rangle$. Using these facts, one sees that r_n can be written more simply as

$$r_n = \begin{cases} \langle 1, 1, \ldots, 1 \rangle & \text{if } \sum m_i \text{ is even} \\ \langle 2, 1, \ldots, 1 \rangle & \text{if } \sum m_i \text{ is odd.} \end{cases} \tag{31.4}$$

EXAMPLES 31.5. For $n = 1, \ldots, 8$, we have:

$$r_1 = \langle 1 \rangle, \quad r_2 = \langle 2 \rangle, \quad r_3 = \langle 2, 1 \rangle, \quad r_4 = \langle 1 \rangle,$$
$$r_5 = \langle 1, 1 \rangle, \quad r_6 = \langle 2, 1 \rangle, \quad r_7 = \langle 2, 1, 1 \rangle, \quad r_8 = \langle 2 \rangle.$$

Recall that a quadratic form q is said to *contain* a quadratic form q' if q' is isomorphic to a subform of q.

PROPOSITION 31.6. [Se 84, Prop. 4] *The trace form q_E contains r_n.*

PROOF. By a well-known theorem of Springer [Lam 73, VII.2.3], it is sufficient to prove this after making an odd-degree extension of k. We then apply Proposition 31.2, together with the obvious fact that the trace form of E'_i contains $\langle 2^{m_i} \rangle$. □

COROLLARY 31.7. *For a quadratic form q of rank n to be a trace form, it is necessary that q contains r_n.*

In what follows, this necessary condition will be called the *trivial* one. For $n > 3$, there are other conditions, stemming from Th. 25.13 (vanishing of the higher Stiefel-Whitney classes of q) and from Th. 29.4 (linear relations between the exterior powers of q).

REMARK 31.8. The subform r_n is the "largest fixed part" of q_E, in the following sense. Fix a ground field k_0 as usual. Write q_E as $r_n \oplus q'_E$, where q'_E has rank $n-h$. Then the "map" $E \mapsto q'_E$ has no fixed part, i.e., there is no $x \in k_0^*$ which is represented by every q'_E (for every k/k_0 and every $E \in \mathrm{Et}_n(k)$).

To prove this, one decomposes n as above into sums of terms 2^{m_i} and chooses for each i an i-Pfister form p_i. Define $q_i = \langle 2^{m_i} \rangle \cdot p_i$; the form $q = \oplus q_i$ is a trace form of rank n. (To check this, one may assume that n is a power of 2, say $n = 2^m$, then a tensor product of m suitably chosen quadratic algebras has the desired trace form.) One then uses the fact that the pure part of a generic Pfister form does not represent any element of k_0, as follows for instance from Prop. 28.1.

31.9. THE CASES $n = 1, 2, 3$. In these cases, the trivial condition is sufficient. In other words:

THEOREM 31.10. (1) *If $n = 1$, q is a trace form if and only if $q \cong \langle 1 \rangle$.*

(2) *If $n = 2$, q is a trace form if and only if it contains $\langle 2 \rangle$, i.e., if and only if $q \cong \langle 2, t \rangle$ for some $t \in k^*$.*

(3) *If $n = 3$, q is a trace form if and only if it contains $\langle 2, 1 \rangle$, i.e., if and only if $q \cong \langle 2, 1, t \rangle$ for some $t \in k^*$.*

PROOF. The necessity follows from Corollary 31.7. The sufficiency is clear for $n = 1$. For $n = 2$, if $q \cong \langle 2, t \rangle$, we choose for E the quadratic algebra $k[X]/(x^2 - a)$, with $a = 2t$; its trace form is $\langle 2, 2a \rangle = \langle 2, t \rangle \cong q$. For $n = 3$, if $q \cong \langle 2, 1, t \rangle$, we choose for E the product of k and the quadratic algebra $k[X]/(x^2 - 2t)$ as above. □

31.11. SPECIAL FORMS OF RANK 3. In order to handle the cases $n = 4, 5, 6, 7$, we need to study some particular quadratic forms f of rank 3. Let $d = d(f)$ be the discriminant of f.

PROPOSITION 31.12. *The following properties are equivalent:*

(1) *$\langle d \rangle \oplus f$ contains $\langle 1 \rangle$, i.e., $\langle d \rangle \oplus f$ represents 1.*

(1′) *$\langle d \rangle \oplus f$ is a 2-Pfister form.*

(1″) *d is represented by $\langle 1 \rangle \oplus \langle d \rangle \cdot f$.*

(2) *$\langle 1 \rangle \oplus \langle d \rangle \cdot f \cong \langle d \rangle \oplus f$.*

(2′) *$\lambda^3(f) - \lambda^2(f) + \lambda^1(f) - \lambda^0(f) = 0$ in $WGr(k)$.*

(3) *There exists $a \in k^*$ which is represented by f and is such that $(a) \cdot (d) = 0$ in $H^2(k)$.*

PROOF. The equivalence of (1) $\Leftrightarrow$ (1′) follows from the fact that a form of rank 4 is a Pfister form if and only if it represents 1 and its discriminant is 1. The equivalence (1) $\Leftrightarrow$ (1″) is obvious (multiplication by $\langle d \rangle$).

(2) $\Rightarrow$ (1) is clear. If (1′) is true, then the Pfister form $\langle d \rangle \oplus f$ represents $\langle d \rangle$, hence is isomorphic to $\langle d \rangle \cdot (\langle d \rangle \oplus f) \cong \langle 1 \rangle \oplus \langle d \rangle \cdot f$. Hence (2) is equivalent to (1), (1′), and (1″).

We have $\lambda^3(f) = \langle d\rangle$, $\lambda^2(f) = \langle d\rangle\cdot f$, $\lambda^1(f) = f$, and $\lambda^0(f) = \langle 1\rangle$. Hence (2′) is equivalent to (2).

We now show (1) $\Leftrightarrow$ (3), which will conclude the proof.

(1) $\Rightarrow$ (3): By (1), we have an equation $1 = dx^2 + a$, where a is represented by f. If a is not 0, we have

$$(a)\cdot(d) = (a)\cdot(dx^2) = (a)\cdot(1-a) = 0,$$

hence (3). If $a = 0$, then d is a square and (3) is obviously true.

(3) $\Rightarrow$ (1): If a is as in (3), the fact that $(a)\cdot(d) = 0$ means that a may be written as $u^2 - d\cdot v^2$ with u, v in k. This shows that $u^2 = d\cdot v^2 + a$ is represented by $\langle d\rangle \oplus f$. If $u \neq 0$, this implies that 1 is represented by $\langle d\rangle \oplus f$; if $u = 0$, then $\langle d\rangle \oplus f$ is isotropic, hence represents 1. □

A quadratic form f of rank 3 with properties (1) through (3) above will be called *special*. Note that, over a finite field, every f is special: use, for instance, (1) or (3). (The same is true over any field k such that $\mathrm{cd}_2(k) \leq 2$, see Prop. 31.14 below.)

REMARK 31.13. If f becomes special over an odd-degree extension of k, then f is special. This follows, for instance, from condition (1).

Here is a less elementary characterization of special forms:

PROPOSITION 31.14. *Properties* (1), ..., (3) *are equivalent to*:

(4) $w_3(f) = 0$ *in* $H^3(k)$.

PROOF. Let us write f as $\langle a, b, c\rangle$ with $d = abc$. Define

$$q_2 = 1 \oplus \langle d\rangle\cdot f = \langle 1, ab, ac, bc\rangle \quad \text{and} \quad q_3 = \langle 1, -d\rangle\cdot q_2.$$

These forms are Pfister forms. The Elman-Lam invariant $e_2(q_2)$ as in §18 is given by:

$$\begin{aligned} e_2(q_2) &= (-ab)\cdot(-ac) \\ &= (a)\cdot(b) + (b)\cdot(c) + (c)\cdot(a) + (-1)\cdot(-abc) \\ &= w_2(f) + (-1)\cdot(-d). \end{aligned}$$

The invariant of q_3 is thus:

$$\begin{aligned} e_3(q_3) &= (d)\cdot e_2(q_2) = (d)\cdot w_2(f) \quad \text{(since } (d)\cdot(-d) = 0\text{)}. \\ &= w_1(f)\cdot w_2(f) = w_3(f), \quad \text{cf. (19.3)}. \end{aligned}$$

By a well-known theorem of Merkurjev (see e.g. [Ar 75, Prop. 2]), q_3 is hyperbolic if and only if its invariant is 0, i.e., if and only if (4) is true. On the other hand, the very definition of q_3 shows that it is hyperbolic if and only if d is represented by q_2, which is condition (1). Hence (1) and (4) are equivalent. □

In what follows, we shall need another characterization of special forms.

PROPOSITION 31.15. *Assume* $|k| > 3$. *Properties* (1), ..., (4) *are equivalent to*:

(5) *There exist* A, B *in* k^* *with* $A^2 - B \neq 0$ *such that*

$$f \cong \langle A, A^2 - B, AB(A^2 - B)\rangle.$$

PROOF. (3) $\Rightarrow$ (5): If k is finite, put $B = d$, and choose $A \in k^*$ such that $A^2 - B \neq 0$. (This is possible since $|k| > 3$.) The form $\langle A, A^2 - B, AB(A^2 - B)\rangle$ has the same discriminant as f, hence is isomorphic to f. If k is infinite, write f as $\langle a, b, c\rangle$ with $d = abc$ and $(a)\cdot(d) = 0$, hence $a = u^2 - d\cdot v^2$ with $u, v \in k$. Since k is infinite, one may choose u, v both nonzero. (Otherwise, the corresponding conic in the (u, v)-plane would have at most 4 points.) Define A, $B \in k^*$ by

$$A = b, \qquad B = d\cdot(bv/u)^2.$$

We have $A^2 - B = b^2\cdot(1 - d\cdot v^2/u^2) = b^2\cdot a/u^2$, hence $\langle A^2 - B\rangle \cong \langle a\rangle$ and $\langle A, A^2 - B, AB(A^2 - B)\rangle \cong \langle b, a, bda\rangle \cong \langle b, a, c\rangle \cong f$.

(5) $\Rightarrow$ (1): If A, B are as in (5), we have $\langle d\rangle = \langle B\rangle$, and the form $\langle d\rangle \oplus f$ contains $\langle B, A^2 - B\rangle$, which represents 1. $\square$

REMARK 31.16. If $|k| = 3$, the form $\langle 1, 1, 1\rangle$ is special, but does not have property (5).

31.17. THE CASE $n = 4$. We now return to the question "*When is q a trace form?*" in the case where q has rank $n = 4$. In that case, the trivial condition is that q represents 1, i.e., q can be written as $\langle 1\rangle \oplus f$, with rank $f = 3$.

THEOREM 31.18 ([Se 94] and [EK 94]). *For $q = \langle 1\rangle \oplus f$ to be a trace form, it is necessary and sufficient that f be special* (in the sense defined in §31.12 above).

PROOF. Necessity: If $q = \langle 1\rangle \oplus f$ is a trace form, we know by Th. 29.4 that $\lambda^3(f - \langle 1\rangle) = 0$. But it is easy to see that

$$\lambda^3(f - \langle 1\rangle) = \lambda^3(f) - \lambda^2(f) + \lambda^1(f) - \lambda^0(f).$$

Hence f has property (2′) of Prop. 31.12, which means that f is special.

[Alternately, Th. 25.13 shows that $w_3(q) = 0$, hence $w_3(f) = 0$, and one applies Prop. 31.14.]

Sufficiency: If k is finite, every quadratic form of rank > 1 is a trace form (just choose $E = k \times \cdots \times k$ or $E = k \times \cdots k \times k_2$ where k_2 is a quadratic extension of k). When k is infinite, we may apply Prop. 31.15 and write f as $\langle A, A^2 - B, AB(A^2 - B)\rangle$. Let P be the *polynôme bicarré*

$$P(X) = X^4 - 2AX^2 + B.$$

Let $E(A, B)$ be the rank 4 algebra $k[X]/(P)$; the following lemma shows that it has the required trace form. $\square$

LEMMA 31.19. *The algebra $E(A, B)$ is étale, and its trace form is isomorphic to $\langle 1, A, A^2 - B, AB(A^2 - B)\rangle$.*

PROOF. The discriminant of P is $64\cdot B\cdot(A^2 - B)^2$, which is not 0. Hence $E(A, B)$ is étale. If x denotes the image of X in $E(A, B)$, one checks that the elements

$$1, \quad x, \quad x^2 - A, \quad x - A/x$$

make up an orthogonal basis of $E(A, B)$, and that the traces of their squares are respectively

$$4, \quad 4A, \quad 4(A^2 - B), \quad 4A(A^2 - B)/B.$$

Hence the trace form of $E(A, B)$ is

$$\langle 4, 4A, 4(A^2 - B), 4A(A^2 - B)/B\rangle \cong \langle 1, A, A^2 - B, AB(A^2 - B)\rangle.$$

□

EXERCISE 31.20. Show that the trace form of $E(A, B)$ is isomorphic to

$$\langle 1, AB, A^2 - B, A(A^2 - B)\rangle.$$

31.21. THE CASE $n = 5$. Here the trivial condition on q is that it contains $\langle 1, 1\rangle$, i.e., that $q = \langle 1, 1\rangle \oplus f$, with $\operatorname{rank} f = 3$.

THEOREM 31.22. *For q to be a trace form, it is necessary and sufficient that f be special.*

PROOF. Necessity: If q is isomorphic to a trace form q_E, there is (Prop. 31.2) an odd-degree extension k'/k over which E splits as a product of k' and of a rank 4 étale algebra. By 31.18, this implies that f becomes special over k', hence is special over k (cf. Remark 31.13).

Sufficiency: By Theorem 31.18 there is an étale algebra E' of rank 4 whose trace form is isomorphic to $\langle 1\rangle \oplus f$. The algebra $E = k \times E'$ has trace form q. □

We could also have deduced Theorem 31.22 directly from Theorem 31.18 and from the following general result:

THEOREM 31.23. *If E is an étale algebra of odd rank n, there is an étale algebra E' of rank $n - 1$ such that E and $k \times E'$ have the same trace form.*

In characteristic 0, this can be deduced from [Mes 90, Prop. 1 and Prop. 3], cf. [EK 94, p. 425]. The result also holds in characteristic $p > 2$ (Mestre, unpublished); we omit the proof.

COROLLARY 31.24. *Let q be a quadratic form of odd rank n. Then q is a trace form if and only if q is isomorphic to $\langle 1\rangle \oplus q'$, where q' is a trace form of rank $n-1$.*

31.25. THE CASE $n = 6$. Let q be a quadratic form of rank 6; let $d = d(q)$ be its discriminant.

THEOREM 31.26. *For q to be a trace form, it is necessary and sufficient that the following two conditions be fulfilled:*

(a) *q contains $\langle 2, 1\rangle$;*

(b) *q contains $\langle 2, 1, 1\rangle$ over the field $k(\sqrt{2d})$.*

PROOF. Necessity: Condition (a) is the trivial condition in rank 6. We may then write q as $\langle 2, 1\rangle \oplus Q$, where $\operatorname{rank}(Q) = 4$, and $d(Q) = 2d$. To prove (b), we may replace k by $k(\sqrt{2d})$, i.e., assume that $2d$ is a square, i.e., $d(Q) = 1$. We may also replace k by any odd-degree extension. By Proposition 31.2, this allows us to assume that q is isomorphic to q_E, where E splits as $E' \times E''$ with $\operatorname{rank} E' = 2$ and $\operatorname{rank} E'' = 4$. We thus have

$$q \cong q_{E'} \oplus q_{E''} \cong \langle 2, 2x\rangle \oplus \langle 1\rangle \oplus f, \text{ with } \operatorname{rank}(f) = 3.$$

Hence $Q \cong \langle 2x\rangle \oplus f$. Since $d(Q) = 1$, we have $\langle 2x\rangle \cong \langle d(f)\rangle$, hence $Q \cong \langle d(f)\rangle \oplus f$. By Theorem 31.18, f is special, hence has property (1) of Proposition 31.12; this means that Q contains $\langle 1\rangle$, hence q contains $\langle 2, 1, 1\rangle$.

Sufficiency: We use the following lemma:

LEMMA 31.27. *Let Q be a quadratic form of rank > 1, and let a be an element of k^*. Suppose that Q represents 1 over the field $k(\sqrt{a})$. Then Q contains a binary form $\langle x, y\rangle$ with $(x)\cdot(ay) = 0$ in $H^2(k)$.*

Assume this lemma. If we write q as $\langle 2,1\rangle \oplus Q$, we may apply the lemma with $a = 2d$, hence $Q \cong \langle x, y\rangle \oplus g$ with $(x)\cdot(2dy) = 0$ and $\mathrm{rank}(g) = 2$. We have $q \cong q' \oplus q''$ with $q' = \langle 2, y\rangle$ and $q'' = \langle 1, x\rangle \oplus g$. The form q' is the trace form of an étale algebra E'. The form q'' can be written as $\langle 1\rangle \oplus f$ where $f = \langle x\rangle \oplus g$. Note that $\langle d(f)\rangle \cong \langle 2dy\rangle$. Hence the relation $(x)\cdot(2dy) = 0$ can be rewritten as $(x)\cdot(d(f)) = 0$. This shows that f has property (3) of Prop. 31.12, hence is special. By Theorem 31.18, there exists an étale algebra E'' of rank 4 whose trace form is q''. Hence q is isomorphic to the trace form of $E' \times E''$. □

PROOF OF LEMMA 31.27. If a is a square, Q represents 1, hence contains $\langle 1, y\rangle$ for some y, and we have $(1)\cdot(ay) = 0$. Assume a is not a square; put $k_1 = k(u)$ with $u^2 = a$. Write V for the k-vector space underlying Q. The assumption that Q represents 1 over k_1 means that there is an element z of $k_1 \otimes_k V$ with $Q(z) = 1$. If we write z as $z = z' + u\cdot z''$, with z', z'' in V, the fact that $Q(z) = 1$ means that z' and z'' are orthogonal and that

$$1 = Q(z') + a\cdot Q(z''). \tag{31.28}$$

Put $x = Q(z')$ and $y = Q(z'')$. If they are both not 0, Q contains $\langle x, y\rangle$ and (31.28) shows that $(ay) = (1 - x)$, hence $(x)\cdot(ay) = 0$, as required. If $y = 0$, (31.28) shows that Q represents 1, hence contains $\langle 1, b\rangle$ for some $b \in k^*$, and we win as above. If $x = 0$, then Q represents a, and then contains $\langle b, a\rangle$ for some $b \in k^*$, and we also win. □

31.29. THE CASE $n = 7$. Let q be a quadratic form of rank 7, and let d be its discriminant.

THEOREM 31.30. *For q to be a trace form, it is necessary and sufficient that:*

(a) *q contains $\langle 2, 1, 1\rangle$;*
(b) *q contains $\langle 2, 1, 1, 1\rangle$ over the field $k(\sqrt{2d})$.*

PROOF. This follows from Cor. 31.24 combined with Theorem 31.26. A direct proof can be given by the method we followed in the case $n = 5$, namely: If q is a trace form q_E, we make an odd-degree extension k'/k over which E splits as $E = k \times E'$ with $\mathrm{rank}\, E' = 6$. By applying Theorem 31.26 to E', we see that q has properties (a) and (b) over k', hence also over k.

Conversely, if q has properties (a) and (b), we write it as $q \cong \langle 1\rangle \oplus q'$, and use Theorem 31.26 to show that there exists an étale algebra E' such that $q_{E'} \cong q'$. For $E = k \times E'$ we have $q_E = q$. □

31.31. REMARKS AND QUESTIONS.

(1) Let q be a quadratic form of rank n. Suppose q becomes a trace form after an odd-degree extension of k. Is it true that q is a trace form (over k)?

 The criteria we have given above show that this is true for $n < 8$.

(2) Let E be an étale algebra of rank n. Is it true that there exists an étale algebra E° over k with the following two properties:
 (2a) $q_{E^\circ} \cong q_E$;
 (2b) E° is split by a Galois extension of k of order a power of 2 (in other words, the image of the homomorphism $\Gamma_k \to S_n$ associated with E° is a 2-group)?

(Note that (2b) implies that E° is a product of fields whose degrees over k are powers of 2.)

Here also, what we have done shows that this is true for $n < 8$. One may be even more optimistic, and guess that there is a higher rank version of Mestre's theory [Mes 90] which changes E into E° by a trace-form preserving deformation. This is known for $n < 4$, but not (yet) for $n = 4$.

(3) What about $n = 8$? Some necessary conditions for a quadratic form q of rank 8 to be a trace form are:
   (i) $q \cong \langle 2\rangle\cdot(\langle 1\rangle \oplus Q)$ for some Q of rank 7, see 31.7.
   (ii) $w_i(Q) = 0$ for $i \geq 5$, see 25.6.
   (iii) $\lambda^i(Q - 3) = 0$ for $i \geq 5$, see 29.4.
   (iv) If $d = d(q) = d(Q)$, the form $Q \oplus \langle d\rangle$ represents 1.

To see that (iv) is necessary, one makes an odd-degree extension so that the Galois group of k is a 2-group. In that case, we have a filtration $E = E_8 \supset E_4 \supset E_2 \supset E_1 = k$, where the subscript is the rank and each algebra is étale of rank 2 over the next one. We may write E_2 as a quadratic algebra $k[X]/(X^2 - a)$. Hence q contains $\langle 2\rangle\cdot\langle 1, a\rangle$, i.e., Q contains $\langle a\rangle$; write Q as $\langle a\rangle \oplus Q'$. The discriminant d is a norm of an element of E_2 (namely, the norm of the discriminant of E/E_2), hence is such that

$$(a)\cdot(d) = 0 \quad \text{in } H^2(k),$$

which is equivalent to $\langle a, d\rangle = \langle 1, ad\rangle$. We then have $Q\oplus\langle d\rangle = Q'\oplus\langle a, d\rangle = Q' \oplus \langle 1, ad\rangle$, which shows that $Q \oplus \langle d\rangle$ represents 1.

It is doubtful that these conditions are sufficient.

32. Trace forms with vanishing Stiefel-Whitney classes

In this section, we consider étale algebras E of rank n which have the same trace form as the split algebra $k \times \cdots \times k$, i.e., such that $q_E \cong \langle 1, \ldots, 1\rangle$.

A necessary condition for this is the vanishing of the Stiefel-Whitney classes $w_i(q_E)$ for all $i > 0$. This is sufficient for $n < 6$. More precisely:

PROPOSITION 32.1. *If $n < 6$ and $w_i(q_E) = 0$ for $i = 1, 2$, then $q_E \cong \langle 1, \ldots, 1\rangle$.*

PROOF. If $n = 5$, by 31.6 we have $q_E \cong \langle 1, 1\rangle \oplus f$ with $\mathrm{rank}(f) = 3$. The form f has the same w_1 and w_2 as the unit form $\langle 1, 1, 1\rangle$, hence is isomorphic to it. This shows that $q_E \cong \langle 1, 1, 1, 1, 1\rangle$.

The case $n < 5$ follows by replacing E by $k \times \cdots \times k \times E$, so that it becomes of rank 5. □

32.2. THE CASE $n = 6$. We now assume that $n = 6$. (The case $n = 7$ is similar, see 32.28.) Note that, in that case, the properties $w_1(q_E) = 0$ and $w_2(q_E) = 0$ imply that $w_3(q_E) = 0$ since $w_3 = w_1\cdot w_2$, and they also imply $w_i(q_E) = 0$ for $i > 3$ by Th. 25.13. Hence the vanishing of the $w_i(q_E)$ for $i > 0$ can be restated as:

$$w_1(q_E) = 0 \quad \text{and} \quad w_2(q_E) = 0, \tag{32.3}$$

or equivalently (cf. Th. 25.10):

$$w_i^{\mathrm{gal}}(E) = 0 \quad \text{for } i = 1, 2. \tag{32.4}$$

THEOREM 32.5. *Let q be a quadratic form of rank 6 such that $w_1(q) = 0$ and $w_2(q) = 0$. For q to be a trace form, it is necessary and sufficient that:*

(1) *q contains $\langle 2, 1\rangle$;*
(2) *$q \cong \langle 1, 1, c, c, c, c\rangle$ for some $c \in k^*$;*
(3) *q becomes isomorphic to $\langle 1, 1, 1, 1, 1, 1\rangle$ over $k(\sqrt{2})$.*

First we note two lemmas.

LEMMA 32.6. *Every quadratic form ψ of rank 6 with*

$$w_1(\psi) = (-1) \text{ and } w_2(\psi) = (-1)\cdot(-1) + (a)\cdot(b) \text{ for some } a,\, b \in k^*$$

is isotropic.

PROOF. This can be found in Pfister [Pf66, p. 123, 3′]. Pfister formulates the hypotheses in terms of invariants "$d(\psi)$" and "$c(\psi)$"; these invariants are related to ours by

$$d(\psi) = (-1) + w_1(\psi) \quad \text{and} \quad c(\psi) = (-1)\cdot(-1) + w_2(\psi),$$

so that our hypotheses mean $d(\psi) = 0$ and $c(\psi) = (a)\cdot(b)$. □

LEMMA 32.7. *Let f be a quadratic form of rank 5 such that $w_1(f) = 0$ and $w_2(f) = 0$. Then $f \cong \langle 1, c, c, c, c\rangle$ for some $c \in k^*$.*

PROOF. Let ψ be the form $\langle -1\rangle \oplus f$. One has $w_1(\psi) = (-1)$ and $w_2(\psi) = 0$. This implies that ψ is isotropic by Lemma 32.6 applied with $a = b = -1$.

Hence f represents 1, and we may write it as $\langle 1\rangle \oplus g$. Let c be any element of k^* represented by g. We may write g as $\langle c\rangle\cdot(\langle 1\rangle \oplus h)$ with $\mathrm{rank}(h) = 3$. One has $w_1(h) = 0$ and $w_2(h) = 0$, hence h is isomorphic to $\langle 1, 1, 1\rangle$. This shows that g is isomorphic to $\langle c, c, c, c\rangle$. □

PROOF OF THEOREM 32.5. Necessity: If $q = q_E$, (1) is the "trivial condition" of 31.7. We may then write q as $\langle 1\rangle \oplus q'$ with $\mathrm{rank}(q') = 5$. Lemma 32.7 applied to $f = q'$ shows that q has property (2).

To prove property (3), we may assume that 2 is a square in k. We may write $q \cong \langle 1, 1\rangle \oplus \langle c\rangle\cdot\langle 1, 1, 1, 1\rangle$. By Th. 31.26, q contains $\langle 2, 1, 1\rangle \cong \langle 1, 1, 1\rangle$, i.e., $\langle c\rangle\cdot\langle 1, 1, 1, 1\rangle$ represents 1, and hence c is a sum of 4 squares. This gives (3).

Sufficiency: By Th. 31.26, we only have to check that q contains $\langle 2, 1, 1\rangle$ over $k(\sqrt{2})$, which is obvious from (3). □

32.8. A REFORMULATION OF THEOREM 32.5. Let us write k_4^* for the subgroup of k^* made up of elements which are represented by the 2-Pfister form $\langle 1, 1, 1, 1\rangle$, i.e., which are sums of four squares. We have $\langle c, c, c, c\rangle \cong \langle c', c', c', c'\rangle$ if and only if c' lies in $c\cdot k_4^*$.

LEMMA 32.9. *$f = \langle 1, c, c, c, c\rangle$ represents 2 if and only if there exists $c' \in c\cdot k_4^*$ with $(2)\cdot(c') = 0$ in $H^2(k)$.*

PROOF. If f represents 2, we may write 2 as $x^2 + cz$, with either $z = 0$ or $z \in k_4^*$. In the first case, 2 is a square in k, and we take $c' = c$. In the second case, we take $c' = cz$. Since $(2)\cdot(-1) = 0$, we have

$$(2)\cdot(cz) = (2)\cdot(-cz) = (2)\cdot(x^2 - 2) = 0.$$

Conversely, if $c' \in c \cdot k_4^*$ and $(2)\cdot(c') = 0$, the form $\langle 2, c'\rangle$ is isomorphic to $\langle 1, 2c'\rangle$. Hence

$$\langle 1\rangle \oplus \langle 1, c', c', c', c'\rangle \cong \langle 2, 2, c', c', c', c'\rangle \cong \langle 2, 1, 2c', c', c', c'\rangle,$$

i.e., $\langle 1, c', c', c', c'\rangle \cong \langle 2, 2c', c', c', c'\rangle$, which shows that $\langle 1, c', c', c', c'\rangle$ represents 2; since this form is isomorphic to f, we see that f represents 2. □

This lemma allows us to rewrite Theorem 32.5 in the following way:

THEOREM 32.10. *Let q be a quadratic form of rank 6 with $w_1(q) = 0$ and $w_2(q) = 0$. For q to be a trace form, it is necessary and sufficient that there exists $c \in k^*$ with the following properties:*

(1) $(2)\cdot(c) = 0$ *in* $H^2(k)$;
(2) $q \cong \langle 1, 1, c, c, c, c\rangle$;
(3) c *is a sum of* 4 *squares in* $k(\sqrt{2})$.

Indeed we have

$$32.5.1 \text{ and } 32.5.2 \Longleftrightarrow 32.10.1 \text{ and } 32.10.2$$

by Lemma 32.9, and 32.10.3 is equivalent to 32.5.3.

32.11. PROPERTY O_6. Note that $q = \langle 1, 1, c, c, c, c\rangle$ is isomorphic to the unit form $\langle 1, 1, 1, 1, 1, 1\rangle$ if and only if c is a sum of 4 squares. Our original question about trace forms can then be translated into the following:

(32.12) *Is it true that* 32.10.1 *and* 32.10.3 *imply that* c *is a sum of* 4 *squares in* k*?*

If it is so (for a given field k), we say that k *has property* O_6. In other words, k has property O_6 if and only if the only trace form of rank 6 over k with vanishing Stiefel-Whitney classes is the unit form $\langle 1, 1, 1, 1, 1, 1\rangle$.

REMARK 32.13. For $c \in k^*$, the following are equivalent:

(i) c is a sum of 4 squares.
(ii) The 3-Pfister form $\langle 1, -c\rangle\cdot\langle 1, 1, 1, 1\rangle$ is hyperbolic.
(iii) One has $(c)\cdot(-1)\cdot(-1) = 0$ in $H^3(k)$.

((i) ⇔ (ii) is standard; (ii) ⇔ (iii) follows from the theorem of Merkurjev quoted in 31.14.)

Moreover, if we multiply c by a sum of 4 squares, the element $(c)\cdot(-1)\cdot(-1)$ does not change. It is thus an *invariant* of the form $q = \langle 1, 1, c, c, c, c\rangle$, which we shall denote by $e(q)$; this invariant will play an essential role in §33.

32.14. EXAMPLES OF FIELDS WHICH HAVE PROPERTY O_6.

32.15. Any field for which $k_4^* = k^*$ has property O_6. This applies if $\mathrm{cd}_2(k) \leq 2$ or if $(-1)\cdot(-1) = 0$ in $H^2(k)$ (since in that case, the Pfister form $\langle 1, 1, 1, 1\rangle$ is hyperbolic). In particular, a field of nonzero characteristic has property O_6.

32.16. A $\mathfrak{p}$-adic field (with finite residue field) has property O_6. This follows from 32.15.

32.17. A number field has property O_6. Indeed, if c has property 32.10.3, it is totally positive in $k(\sqrt{2})$, hence also in k, and it is well-known that this implies that c is a sum of 4 squares.

In order to give further examples, we need the following result:

LEMMA 32.18. *Let K be a field with a discrete valuation v, with residue field k of characteristic $\neq 2$. Let $c \in K^*$ be such that $(2)\cdot(c) = 0$ in $H^2(K)$ and that c is a sum of 4 squares in $K(\sqrt{2})$. Then the residue of $(c)\cdot(-1)\cdot(-1) \in H^3(K)$ is 0.*

(Recall that this residue belongs to $H^2(k)$, see 7.13.)

PROOF. We may assume that K is complete. Put $m = v(c)$. The residue of $(c)\cdot(-1)\cdot(-1)$ is equal to 0 if m is even, and to $(-1)\cdot(-1)$ if m is odd. In the second case, the residue of $(2)\cdot(c)$ is (2). Since $(2)\cdot(c) = 0$, this means that 2 is a square in k, hence also in K, and we have $K(\sqrt{2}) = K$, hence c is a sum of 4 squares in K, and we have $(c)\cdot(-1)\cdot(-1) = 0$. □

THEOREM 32.19. *If k has property O_6, the same is true for every purely transcendental extension K of k.*

PROOF. It is enough to prove this for $K = k(T)$, where T is an indeterminate. Let $c \in K^*$ have properties 32.10.1 and 32.10.3, and let $e \in H^3(K)$ be the element $(c)\cdot(-1)\cdot(-1)$. By the lemma above, all the residues of e are 0. This means that e is unramified in the sense of 9.4, hence constant. To prove that it is 0, it is enough to show that the value of e at a suitable v of degree 1 is 0. To do so, choose v with $k(v) = k$ and $v(c) = 0$. (Such a v exists since we may assume by 32.15 that the characteristic of k is 0, hence that k is infinite.) If c' denotes the image of c in $k(v)^* = k^*$, the value of e at v is $(c')\cdot(-1)\cdot(-1)$, where the cup product is computed in $H(k)$. Since $(2)\cdot(c) = 0$, we have $(2)\cdot(c') = 0$, and since $(c)\cdot(-1)\cdot(-1) = 0$ in $H^3(K(\sqrt{2}))$, we have $(c')\cdot(-1)\cdot(-1) = 0$ in $H^3(k(\sqrt{2}))$. Since k has property O_6, these properties imply that c' is a sum of 4 squares in k, hence the value of e at v is 0. □

COROLLARY 32.20. *A purely transcendental extension of a number field has property O_6.*

This follows from 32.17 and 32.19.

EXERCISE 32.21. Prove:

(1) If 2, -1, or -2 is a square in k, then k has property O_6. (Note that this shows that every field of characteristic $\neq 0$ has property O_6.)

(2) Let $\chi_2 : \Gamma_k \to \mathbf{Z}_2^*$ be the 2-adic cyclotomic character. If k does not have property O_6, then χ_2 is surjective. [Hint: Use (1).]

EXERCISE 32.22. Let K be a field which is complete with respect to a discrete valuation, with residue field k of characteristic $\neq 2$. Show that if k has property O_6, then so does K.

[Apply Lemma 32.18 as in the proof of Th. 32.19.]

EXERCISE 32.23. Let q_2 be a 2-Pfister form and let q be a quadratic form of rank 6 with $w_1(q) = 0$ and $w_2(q) = w_2(q_2)$. Show that q is a trace form if and only if there exists $c \in k^*$ such that:

(1) $q \cong \langle 1, 1\rangle \oplus \langle c\rangle \cdot q_2$;

(2) $(2)\cdot(c) = 0$;

(3) c is representable by q_2 over $k(\sqrt{2})$.

[Use the same method as for Th. 32.10, which is the special case $q_2 = \langle 1, 1, 1, 1\rangle$.]

32.24. AN EXAMPLE OF A FIELD WHICH DOES NOT HAVE PROPERTY O_6. Let k be the function field over $\mathbf{Q}$ of the affine quadric

$$X_1^2 + X_2^2 + \cdots + X_5^2 + 7 = 0.$$

PROPOSITION 32.25. *The quadratic form* $q = \langle 1,1,-1,-1,-1,-1\rangle$ *is a trace form over* k *with trivial Stiefel-Whitney classes; it is not isomorphic to the unit form* $\langle 1,1,1,1,1,1\rangle$.

COROLLARY 32.26. *The field* k *does not have property* O_6.

PROOF OF PROP. 32.25. To see that q is a trace form, one applies Th. 32.5 with $c = -1$. Condition 32.5.1 is $(2)\cdot(-1) = 0$, which is clear; condition 32.5.2 is obvious. To check 32.5.3, observe that 7 can be written as $u^2 + v^2 + w^2$ over $k_1 = k(\sqrt{2})$ by choosing, e.g., $u = 1$, $v = \sqrt{2}$, and $w = 2$. If x_i denotes the image in k of the indeterminate X_i, we have

$$x_1^2 + \cdots + x_5^2 + u^2 + v^2 + w^2 = 0.$$

Hence the rank 8 form $\psi = \langle 1,1,\ldots,1\rangle$ is isotropic over k_1, which is equivalent to saying that -1 is a sum of 4 squares over k_1.

It remains to prove that -1 is not a sum of 4 squares over k. Let ϕ be the quadratic form $\langle 1,1,1,1,1,7\rangle$ and let $\mathbf{Q}(\phi)$ be the function field of the affine cone $\phi = 0$, i.e., the field of fractions of $\mathbf{Q}[Y_1,\ldots,Y_6]/(Y_1^2+\cdots+Y_5^2+7Y_6^2)$. The field k defined above embeds into $\mathbf{Q}(\phi)$ by $x_i \mapsto Y_i/Y_6$, for $i = 1,\ldots,5$. If -1 is a sum of 4 squares over k, then ψ splits over $\mathbf{Q}(\phi)$. By [Scha 85, p. 155, Th. 5.4(ii)], applied with $K = \mathbf{Q}$ and $\alpha = 1$, this implies that ψ contains ϕ, hence $\langle 1,1,1\rangle$ contains $\langle 7\rangle$. That is, 7 is representable by $\langle 1,1,1\rangle$ over $\mathbf{Q}$, which is not true. Hence -1 is not a sum of 4 squares over k.

(The only property of 7 that we have used is that it is not a sum of 3 squares over $\mathbf{Q}$, but becomes one over $\mathbf{Q}(\sqrt{2})$.) □

QUESTION 32.27. The field k from 32.24 is an extension of $\mathbf{Q}$ of transcendence degree 4. Do there exist extensions of $\mathbf{Q}$ of smaller transcendence degree (e.g., equal to 1) which fail to have property O_6?

32.28. THE CASE $n = 7$. This case reduces to the case $n = 6$, as explained in §31. In particular:

THEOREM 32.29. *Let* q *be a quadratic form of rank* 7 *over* k, *with* $w_1(q) = 0$ *and* $w_2(q) = 0$. *For* q *to be a trace form, it is necessary and sufficient that:*

(a) q *contains* $\langle 2,1,1\rangle$;
(b) $q \cong \langle 1,1,1,c,c,c,c\rangle$ *for some* $c \in k^*$;
(c) $q \cong \langle 1,1,1,1,1,1,1\rangle$ *over* $k(\sqrt{2})$.

This follows from Theorem 32.5 by using the fact (Cor. 31.24) that q is a trace form if and only if it may be written as $\langle 1\rangle \oplus q'$ where q' is a rank 6 trace form.

Similarly:

THEOREM 32.30. *The following are equivalent:*

(1) k *has property* O_6.
(2) *The only rank* 7 *quadratic form having properties* (a), (b), *and* (c) *of Theorem* 32.29 *is the unit form* $\langle 1,1,1,1,1,1,1\rangle$.

In other words, if one defines "property O_7" in an obvious way, then O_7 is equivalent to O_6.

REMARK 32.31. If q is as in 32.29, we write $e(q)$ for the invariant $(c)\cdot(-1)\cdot(-1)$ in $H^3(k)$ as in 32.13. If k does not have property O_6, there exists an étale algebra E of rank 7 with trivial Stiefel-Whitney classes and $e(q_E) \neq 0$. Moreover, by what has been proved in §31, one may choose for E an algebra of the form

$$E = E_1 \times E_2 \times E_4, \quad \text{with } \operatorname{rank} E_i = i,$$

and with the further property that E_4 is given by a polynôme bicarré as in the proof of Th. 31.18.

33. Application of trace forms to Noether's problem

Until 33.14, we place no restriction on the characteristic of k.

33.1. VERSAL TORSORS AND NOETHER'S PROBLEM: DEFINITIONS. Let G be a finite group. We shall be interested in the following property of G:

$\mathrm{Rat}(G/k_0)$ – *There exist an extension K/k_0 and a G-torsor T over K such that*:

(1) T is versal (*as defined in* 5.1);
(2) K is k_0-rational (*i.e., is a purely transcendental extension of* k_0)

Many of the groups we have considered at the beginning of these notes have that property; for instance the symmetric group S_n, or an elementary abelian group of type $(2, \ldots, 2)$.

A related property is:

$\mathrm{Noe}(G/k_0)$ – *There exists an embedding $\rho : G \to \mathbf{GL}_n(k_0)$ such that, if K_ρ is the subfield of $k_0(X_1, \ldots, X_n)$ fixed by G, then K_ρ is k_0-rational.*

Deciding whether $\mathrm{Noe}(G/k_0)$ is true is "Noether's problem" for G and k_0. If $\mathrm{Noe}(G/k_0)$ is false, we say that *Noether's problem has a negative solution for G over k_0.*

REMARK 33.2. The property $\mathrm{Noe}(G/k_0)$ is known to be equivalent to each of the following:

(1) There exists an embedding ρ for which K_ρ/k_0 is "stably rational" (i.e., becomes rational after adjunction of finitely many indeterminates).
(2) For every embedding ρ, K_ρ/k_0 is stably rational.

THEOREM 33.3. $\mathrm{Noe}(G/k_0)$ *implies* $\mathrm{Rat}(G/k_0)$.

PROOF. Indeed, in the situation of $\mathrm{Noe}(G/k_0)$, the embedding $G \to \mathbf{GL}_n(k_0)$ defines a versal torsor as in 5.4. □

REMARK 33.4. The converse implication is false in general. For instance, if G is cyclic of order 47, then $\mathrm{Noe}(G/\mathbf{Q})$ is not true (Swan [Sw 69], Voskresenskiĭ [Vo 70]), but $\mathrm{Rat}(G/\mathbf{Q})$ is true [Vo 77]. See e.g. Swans' report [Sw 83].

33.5. UNRAMIFIED COHOMOLOGY CLASSES. We need a generalization of the notion of unramified cohomology class defined in 9.4. Consider a finitely generated extension K/k_0; let C be a finite Γ_{k_0}-module.

DEFINITION 33.6. An element $a \in H(K, C)$ is said to be *unramified over* k_0 if, for every discrete valuation v of K which is trivial on k_0, the residue of a at v (as defined in 7.13) is 0.

PROPOSITION 33.7. *If K/k_0 is rational, every unramified cohomology class in $H(K, C)$ is constant, i.e., belongs to $H(k_0, C)$.*

This is a consequence of Th. 10.1.2.

REMARK 33.8. Conversely, every constant cohomology class is obviously unramified.

33.9. UNRAMIFIED COHOMOLOGICAL INVARIANTS. Let G be a finite group, let C be as in 33.5, and let a be an element of $\mathrm{Inv}_{k_0}(G, C)$, i.e., a cohomological invariant of the G-torsors over $\mathrm{Fields}_{/k_0}$ with values in $H(*, C)$.

We say that a is *unramified* if, for every K/k_0 which is finitely generated and every G-torsor T over K, the cohomology class $a(T) \in H(K, C)$ is unramified over k_0 in the sense defined in 33.6.

PROPOSITION 33.10. *If $a \in \mathrm{Inv}_{k_0}(G, C)$ is unramified over k_0, and if $\mathrm{Rat}(G/k_0)$ is true, then a is constant.*

PROOF. Choose a versal G-torsor T over K as in $\mathrm{Rat}(G/k_0)$. By 33.7, $a(T)$ in $H(K, C)$ is constant, i.e., comes from an element $a_0 \in H(k_0, C)$. By replacing a by $a - a_0$, we may assume that $a(T) = 0$. By Th. 12.3, this implies that $a = 0$. □

COROLLARY 33.11. *Suppose that $\mathrm{Rat}(G/k_0)$ is true and that a is normalized* (as defined in 4.5) *and unramified. Then $a = 0$.*

REMARKS 33.12. (1) In what follows, we shall apply 33.11 in the following way: for suitable G and k_0, we shall construct a non-zero cohomological invariant a which is unramified and normalized. By 33.11, this will show that $\mathrm{Rat}(G/k_0)$ is not true, and hence (by 33.3) that *Noether's problem has a negative solution for G over k_0.*

(2) What we are doing here for cohomological invariants could also be done for *Witt invariants.*

REMARK 33.13. The strategy we are using is essentially due to Saltman and Bogomolov. However, there is a difference: Bogomolov and Saltman used mainly cohomological invariants coming from elements of $H(G, C)$ (with trivial action of G on C) as in 4.2; the invariant we shall use will not be of that form (at least not in an obvious way); rather, it will come from the properties of the trace form in rank 6 or 7 proved in §32.

EXAMPLE 33.14 (A cohomological invariant for cyclic groups of order 8, 16, ...). From now until the end of this section, the characteristic is $\neq 2$, and $C = \mathbf{Z}/2\mathbf{Z}$; for k an extension of k_0, we write as usual $H^i(k)$ for $H^i(k, C)$.

Take G cyclic of order 2^m with $m \geq 3$. Every $x \in H^1(k, G)$ defines by reduction (mod 2) an element $d(x)$ of $H^1(k, \mathbf{Z}/2\mathbf{Z}) = H^1(k)$. Put:

$$b(x) = (2) \cdot d(x) \quad \text{in } H^2(k).$$

This defines an invariant b in $\mathrm{Inv}_{k_0}(G, \mathbf{Z}/2\mathbf{Z})$.

PROPOSITION 33.15. *The invariant b defined above is normalized and unramified; if the ground field k_0 is $\mathbf{Q}$, this invariant is not 0.*

PROOF. The invariant b is clearly normalized. Let us show that it is unramified. Let k be an extension of k_0 which is complete for a discrete valuation which is trivial on k_0, and let I denote the corresponding inertia subgroup of Γ_k. Let x be an element of $H^1(k, G)$; call φ_x the corresponding homomorphism $\Gamma_k \to G$. We have to show that $b(x) = (2) \cdot d(x)$ is an unramified element of $H^2(k)$. If $\varphi_x(I) \neq G$, the cohomology class $d(x)$ is unramified, and so is $b(x)$. If $\varphi_x(I) = G$, the residue field contains the 2^m-th roots of unity; since $m \geq 3$, this implies that 2 is a square in the residue field, hence 2 is a square in k and $b(x) = 0$. This shows that the invariant b is unramified over k_0.

Take k' to be the unramified extension of degree 2^m of $k = \mathbf{Q}_2$ (2-adic field). Then k'/k defines an element x of $H^1(k, G)$ such that $d(x)$ is equal to 5 (in k^*/k^{*2}). Since the quaternion k-algebra $(2, 5)$ is a division algebra, we have $(2) \cdot (5) \neq 0$ in $H^2(k)$, which shows that $b(x)$ is not 0. Hence b is a nonzero element of $\mathrm{Inv}_{\mathbf{Q}}(G, \mathbf{Z}/2\mathbf{Z})$. □

THEOREM 33.16. *Let G be a group with a 2-Sylow subgroup which is cyclic of order ≥ 8. Then* $\mathrm{Rat}(G/\mathbf{Q})$ — *and a fortiori* $\mathrm{Noe}(G/\mathbf{Q})$ — *are false.*

PROOF. Let H denote a 2-Sylow subgroup of G. Since it is cyclic, it has a normal complement N [Sco 87, 6.2.11]. The invariant b defined above gives an invariant a of G via the map $G \to H$. This invariant is normalized and unramified since b is. Let k be an extension of $\mathbf{Q}$ and let t be an element of $H^1(k, H)$ such that $b(t)$ is nonzero. Since the map $H^1(k, G) \to H^1(k, H)$ is surjective, a is not 0 and we may apply 33.11. □

REMARK 33.17. The special case of Th. 33.16 where G is abelian was proved by:

- Endo-Miyata [EM 73], Voskresenskiĭ [Vo 73], and Lenstra [Le 74] (for the Noe property);
- Voskresenskiĭ [Vo 77] and Saltman [Sal 82] (for the Rat property).

Saltman's proof uses a cohomological invariant with values in $H^2(k)$ which — first appearances to the contrary — is the same as the invariant b defined in Example 33.14 above.

33.18. THE GROUPS $\widetilde{A}_n$ AND THE CORRESPONDING ÉTALE ALGEBRAS. For every $n \geq 4$, we write $\widetilde{A}_n$ for the inverse image of A_n in the extension $\widetilde{S}_n \to S_n$ from 25.3; this is the unique non-trivial extension of A_n by $\{\pm 1\}$. We have an exact sequence

$$1 \to \{\pm 1\} \to \widetilde{A}_n \to A_n \to 1. \tag{33.19}$$

Consider now an étale algebra E of rank n over a field k, and let

$$\varphi_E \colon \Gamma_k \to S_n$$

be the corresponding homomorphism as in 3.2. Let $w_i(q_E)$ be the Stiefel-Whitney classes of the trace form q_E. By 25.3 and 25.10, we have:

(33.20) The image of φ_E is contained in A_n if and only if $w_1(q_E) = 0$.

(33.21) Suppose that the image of φ_E is contained in A_n. The homomorphism $\varphi_E \colon \Gamma_k \to A_n$ can be lifted to $\widetilde{A}_n$ if and only if $w_2(q_E) = 0$.

In particular:

PROPOSITION 33.22. *The image of the map*

$$H^1(k, \widetilde{A_n}) \to H^1(k, A_n) \to H^1(k, S_n) = \mathrm{Et}_n(k)$$

is equal to the classes of rank n étale algebras E with $w_1(q_E) = 0$ and $w_2(q_E) = 0$.

33.23. A COHOMOLOGICAL INVARIANT FOR $\widetilde{A_6}$ AND $\widetilde{A_7}$. Set $n = 6$ or 7, let t be an element of $H^1(k, \widetilde{A_n})$, and let $E(t)$ in $\mathrm{Et}_n(k)$ be the corresponding étale algebra. By 33.22, we may apply the results of §32 to $E(t)$. In particular, the trace form of $E(t)$ is of the form

$$\langle 1, 1, c, c, c, c\rangle \text{ (for } n = 6) \quad \text{or} \quad \langle 1, 1, 1, c, c, c, c\rangle \text{ (for } n = 7)$$

for a suitable $c \in k^*$. Moreover, the element $(c)\cdot(-1)\cdot(-1) \in H^3(k)$ does not depend on the choice of c (as explained in 32.13); let us denote it by $e(t)$. We may view e as a *cohomological invariant*

$$e\colon H^1(k, \widetilde{A_n}) \to H^3(k).$$

PROPOSITION 33.24. *The invariant e defined above is normalized and unramified; if the ground field k_0 is $\mathbf{Q}$, this invariant is not 0.*

PROOF. That e is normalized is clear; that it is unramified follows from Theorem 32.10 and Lemma 32.18. To prove that $e \neq 0$ if $k_0 = \mathbf{Q}$, one needs only to construct an extension $k/\mathbf{Q}$ and an element t of $H^1(k, \widetilde{A_n})$ with $e(t) \neq 0$; this has been done in 32.25, by choosing k to be the function field of the affine quadric $X_1^2 + \cdots + X_5^2 + 7 = 0$. □

THEOREM 33.25. *Noether's problem has a negative solution for $\widetilde{A_6}$ and $\widetilde{A_7}$ over $\mathbf{Q}$.*

This follows from Remark 33.12.1.

The method we have used gives a bit more, namely:

THEOREM 33.26. *Let G be a subgroup of odd index of $\widetilde{A_7}$. Then Noether's problem has a negative solution for G over $\mathbf{Q}$.*

(In fact, we shall prove the stronger result that $\mathrm{Rat}(G/\mathbf{Q})$ is false.)

PROOF. By composition, e defines a cohomological invariant

$$e_G\colon H^1(k, G) \to H^1(k, \widetilde{A_7}) \xrightarrow{e} H^3(k).$$

It is clear that e_G is unramified and normalized. Let us show that it is not 0. Let H be a 2-Sylow subgroup of $\widetilde{A_7}$ contained in G. (It is easy to see that H is isomorphic to the quaternion group Q_{16} of order 16, but this will not play any role here.) By Remark 32.31, there exists an étale algebra E of rank 7 over a suitable field k such that $e(E)$ is not zero and E is split by a Galois extension of k of degree a power of 2. Such an algebra comes from a homomorphism $\Gamma_k \to \widetilde{A_7} \to A_7$ whose image is a 2-group. Hence, it comes from a homomorphism $\Gamma_k \to H$, hence from an element of $H^1(k, G)$. This shows that e_G is not 0. □

EXAMPLE 33.27. Among the subgroups of odd index of $\widetilde{A_7}$, there are the following:

$$\mathbf{SL}_2(\mathbf{F}_9) = \widetilde{A_6}, \quad \mathbf{SL}_2(\mathbf{F}_7), \quad Q_{16}, \quad \widehat{S_4}, \quad \widehat{S_5}.$$

(Here $\widehat{S_n}$ denotes the unique central extension of S_n by $\{\pm 1\}$ in which the transpositions and the product of two disjoint transpositions lift as elements of order 4. This is not the same as the $\widetilde{S_n}$ of 25.3.)

33.28. ANOTHER DEFINITION OF THE INVARIANT e. The definition of e can be extended to any $\widetilde{A_n}$ $(n \geq 6)$ as follows: Consider the unit quadratic form $q_n = \langle 1, 1, \ldots, 1\rangle$ of rank n and the corresponding orthogonal group $\mathbf{O}_n$. There is a natural embedding of S_n in $\mathbf{O}_n(k)$, which maps A_n in $\mathbf{SO}_n(k)$. One checks (see e.g. [Se 84, 2.3]) that A_n is contained in the image of the covering map

$$\mathbf{Spin}_n(k) \to \mathbf{SO}_n(k)$$

and that its inverse image in $\mathbf{Spin}_n(k)$ is $\widetilde{A_n}$. We thus get an embedding

$$\widetilde{A_n} \to \mathbf{Spin}_n(k) \tag{33.29}$$

(which is in fact the construction of $\widetilde{A_n}$ originally given in [Schur 11] for $k = \mathbf{C}$).

Hence any continuous homomorphism

$$\varphi\colon \Gamma_k \to \widetilde{A_n}$$

defines an element (φ) of $H^1(k, \mathbf{Spin}_n)$. By using the Arason invariant (a special case of Rost's invariant)

$$H^1(k, \mathbf{Spin}_n) \to H^3(k),$$

we map (φ) to an element of $H^3(k)$. We thus get a normalized invariant e_n on the $\widetilde{A_n}$-torsors.

(Here is a more direct description of the invariant. For t an element of $H^1(k, \widetilde{A_n})$, let $E(t)$ be the corresponding rank n étale algebra and write q_t for its trace form. By 33.20 and 33.21, we have $w_1(q_t) = 0$ and $w_2(q_t) = 0$. This means that the element $q_t - q_n$ of the Witt ring $W(k)$ belongs to I^3, where I is the augmentation ideal of $W(k)$. The Arason map $I^3/I^4 \to H^3(k)$ transforms $q_t - q_n$ into an element of $H^3(k)$, which is easily seen to equal $e_n(t)$.)

When $n \leq 5$, the invariant e_n is 0, see Prop. 32.1.

When $n = 6, 7$, one checks that e_n coincides with the invariant e defined above.

For $n \geq 8$, it is no longer true that e_n is unramified, as simple examples show; hence, one cannot use it to settle Noether's problem for $\widetilde{A_n}$.

34. Further applications to Noether's Problem

34.1. Set H to be the quaternion group Q_{16} of order 16, i.e., the group generated by two elements r, s with the definining relations

$$r^8 = 1, \quad s^2 = r^4, \quad \text{and} \quad srs^{-1} = r^{-1}.$$

The center $Z(H)$ of H is $\{1, r^4\}$. We identify the group $H/Z(H)$ with the dihedral group D_4 as follows: view D_4 as the subgroup of S_4 which stabilizes the partition $\{1,2\}$, $\{3,4\}$, and map r to the cyclic permutation $(1\,3\,2\,4)$ and s to $(1\,2)$. The quadratic character χ of H defined by $H \to D_4 \to S_4 \to \{1,-1\}$ is given by $\chi(r) = -1$ and $\chi(s) = -1$.

There is a different identification of $H/Z(H)$ with D_4 which gives the character χ' such that $\chi'(r) = -1$ and $\chi'(s) = 1$. Call ψ the product $\chi{\cdot}\chi'$; one has $\psi(r) = 1$ and $\psi(s) = -1$.

In the proof of Th. 33.26, we defined an element e of $\mathrm{Inv}(H, \mathbf{Z}/2\mathbf{Z})$. Let us recall its definition: if k is a field of characteristic not 2, and x belongs to $H^1(k, H)$, then x defines an element x_4 of $H^1(k, D_4)$, hence a rank 4 étale algebra $E_4(x)$; on the other hand, x defines via $\chi\colon H \to \{1, -1\}$ an element of $H^1(k, \{1, -1\})$, hence a quadratic algebra $E_2(x)$. The product

$$E_6(x) = E_4(x) \times E_2(x) \tag{34.2}$$

is a rank 6 étale algebra over k, with trivial w_1 and w_2 (see (33.20) and (33.21)). Hence by Th. 32.5 its trace form $q_x = q_{E_6(x)}$ may be written as $\langle 1, 1, c, c, c, c\rangle$ for some $c \in k^*$. *The invariant* $e(x)$ *is then defined as*

$$e(x) = (-1){\cdot}(-1){\cdot}(c) \quad \textit{in } H^3(k), \tag{34.3}$$

and we have seen that e is a nonzero element of $\mathrm{Inv}_{k_0}(H, \mathbf{Z}/2\mathbf{Z})$ if the ground field k_0 is chosen to be $\mathbf{Q}$.

We now show that the invariant e has property (2) of Prop. 15.9:

PROPOSITION 34.4. $\mathrm{Res}_H^{H'}(e) = 0$ *for every proper subgroup* H' *of* H.

PROOF. We have to show that, if $x \in H^1(k, H)$ corresponds to a homomorphism $\varphi\colon \Gamma_k \to H$ which is not surjective, then $e(x) = 0$, i.e., the trace form q_x is isomorphic to $\langle 1, 1, 1, 1, 1, 1\rangle$. Any proper subgroup of H is contained in a subgroup of index 2, hence in the kernel of χ, χ', or ψ. This gives us three cases to consider. To do so, we write q_x explicitly as in §32: the algebra $E_4(x)$ can be defined by a polynôme bicarré $X^4 - 2AX^2 + B$, hence has trace form $\langle 1, A, A^2 - B, AB(A^2 - B)\rangle$, see Lemma 31.19; the algebra $E_2(x)$ has trace form $\langle 2, 2B\rangle$, so that we have:

$$q_x = \langle 1, 2, A, 2B, A^2 - B, AB(A^2 - B)\rangle.$$

By composing $\varphi\colon \Gamma_k \to H$ with a quadratic character ε of H, one gets a quadratic character of Γ_k, hence an element x_ε of k^*/k^{*2}. One finds:

$$x_\chi = B, \quad x_\psi = B(A^2 - B), \quad \text{and} \quad x_{\chi'} = A^2 - B.$$

We can now handle the three cases:

Case I: $\mathrm{im}(\phi)$ *is contained in* $\ker(\chi)$. This means that B is a square. hence q_x may be written as

$$\langle 1, 2, A, 2, A^2 - B, A(A^2 - B)\rangle,$$

which contains $\langle 1, 2, 2\rangle = \langle 1, 1, 1\rangle$; this implies that "$c$" can be chosen to be equal to 1, and hence $e(x) = 0$.

Case II: $\mathrm{im}(\phi)$ *is contained in* $\ker(\chi')$. This means that $A^2 - B$ is a square, hence q_x contains $\langle 1, 1, 2\rangle$ and "c" can be chosen to be equal to 2, and we have $e(x) = 0$ since 2 is a sum of 4 squares.

Case III: $\mathrm{im}(\phi)$ *is contained in* $\ker(\psi)$. This means that $B(A^2 - B)$ is a square, i.e., we have $A^2 - B = Bt^2$ for some $t \in k^*$. This implies that $B = A^2/(1 + t^2)$, hence B is a sum of two squares and we have $(-1){\cdot}(B) = 0$. The trace form q_x may be written

$$q_x = \langle 1, 2, A, 2B, B, A\rangle = \langle 1, 2, B\rangle \oplus \langle A, A, 2B\rangle.$$

We have

$$w_2(q_x) = (-1)\cdot(A) + (2)\cdot(B).$$

Since $w_2(q_x) = 0$, this implies that $(-1)\cdot(A) = (2)\cdot(B)$. The two quadratic forms $\langle 1, 2, B\rangle$ and $\langle A, A, 2B\rangle$ have the same w_1 and w_2, hence are isomorphic. This allows us to rewrite q_x as

$$q_x = \langle 1, 2, B\rangle \oplus \langle 1, 2, B\rangle = \langle 1, 1, 2, 2, B, B\rangle.$$

Since 2 and B are sums of two squares, we have

$$\langle 2, 2\rangle = \langle 1, 1\rangle = \langle B, B\rangle,$$

hence $q_x = \langle 1, 1, 1, 1, 1, 1\rangle$ and $e(q_x) = 0$. □

34.5. Now let G be a finite group with a 2-Sylow subgroup H isomorphic to Q_{16}. Let $e \in \mathrm{Inv}(H, \mathbf{Z}/2\mathbf{Z})$ be the invariant defined in (34.3).

PROPOSITION 34.6. *Let $b \in \mathrm{Inv}(G, \mathbf{Z}/2\mathbf{Z})$ be the corestriction of e as defined in* 14.3. *Then:*

(i) *The restriction of b to H is equal to e.*
(ii) *The invariant b is unramified.*
(iii) *We have $b \neq 0$ if the ground field k_0 is $\mathbf{Q}$.*

PROOF. Let N be the normalizer of H in G. The group N/H is of odd order. On the other hand, the group of automorphisms of H is easily seen to be of order 16. This shows that the natural homomorphism $N/H \to \mathrm{Out}(H)$ is trivial. In other words, any automorphism of H of the form $x \mapsto gxg^{-1}$ with $g \in N$ is inner. By Prop. 13.1, this implies that such an automorphism fixes e. Hence e has the first property of Prop. 15.9. By Prop. 34.4, it also has the second property of 15.9. This implies (i).

Property (ii) follows from the analogous property of e, combined with the computation of the residue of a corestriction (see 8.6). As for (iii), it is obvious since $e = \mathrm{Res}(b)$ is nonzero when $k_0 = \mathbf{Q}$, see 33.26 and 32.31. □

THEOREM 34.7. *If G has a 2-Sylow subgroup isomorphic to the quaternion group Q_{16}, then $\mathrm{Rat}(G/\mathbf{Q})$ is false.*

This follows from Prop. 34.6 combined with Cor. 33.11.

COROLLARY 34.8. *Noether's problem has a negative solution for G over $\mathbf{Q}$.*

Examples of such groups G are $\mathbf{SL}_2(\mathbf{F}_q)$ with $q \equiv 7$ or $9 \pmod{16}$. (Note that the cases $q = 7$ and $q = 9$ have already been treated in 33.27. Note also that the results above contain as special cases those of Th. 33.26 and 33.27.)

APPENDIX A

A letter from M. Rost to J-P. Serre

Regensburg, den 3.3.91

Dear Professor Serre,

Thank you very much for your letter. Your invariant[a] is interesting — in fact I found it at first somewhat curious and it concerned me some time. As far as I know it is the only cohomological invariant vanishing on isotropic forms. Let me make the following remarks.

I) Relation with exterior algebra

Roughly said, I view your construction as the natural way to give a decomposition of the exterior algebra of a form with respect to the $*$-operator.

To be more precise let me first introduce for an n-dimensional form $\varphi \simeq \langle a_1, \ldots, a_n\rangle$ with $\delta = a_1 \ldots a_n$ the following invariants (taking values say in the set of isometry classes of Pfister forms; hence they yield also cohomological invariants):[b]

$$\begin{aligned}
\alpha(\varphi) &= \langle\!\langle -a_1, \ldots, -a_n \rangle\!\rangle \\
\beta(\varphi) &= \langle\!\langle -a_1, \ldots, -a_{n-1} \rangle\!\rangle \quad \bmod \langle\!\langle \delta \rangle\!\rangle, \quad n \geq 3 \\
\gamma(\varphi) &= \langle\!\langle -a_1a_2, \ldots, -a_1a_{n-1} \rangle\!\rangle \quad \bmod \langle\!\langle \delta \rangle\!\rangle, \quad n \text{ even}
\end{aligned}$$

Here $\langle\!\langle -b_1, \ldots, -b_k \rangle\!\rangle = \langle 1, b_1\rangle \ldots \langle 1, b_k\rangle$. Your invariant is the cohomological version of $\beta(-\varphi)$.

To interpret these invariants it is appropriate to consider the exterior powers $\wedge^k\varphi \colon \wedge^k V \to F$ of a form $\varphi \colon V \to F$. Then

$$\wedge^k\varphi(v_1 \wedge \cdots \wedge v_k) = \varphi(v_1)\cdots\varphi(v_k)$$

for $v_i \perp v_j$ and one finds for $a \in D(\varphi) = \varphi(V) \setminus \{0\}$:

$$\begin{aligned}
\alpha(\varphi) &= \langle\!\langle -\delta \rangle\!\rangle \beta(\varphi) \simeq \wedge\varphi = \oplus_k \wedge^k \varphi \\
\beta(\varphi) &\simeq \wedge^{\text{even}}\varphi = \oplus_k \wedge^{2k} \varphi \quad \text{if } n \text{ is odd}, \delta = 1 \\
\gamma(\varphi) &\simeq \widetilde{\gamma}(\varphi) := \oplus_{k=0}^{\frac{n-2}{4}} \wedge^{2k} \varphi \quad \text{if } n \equiv 2 \pmod 4, \delta = 1 \\
a\gamma(\varphi) &\simeq \widetilde{\gamma}(\varphi) := \oplus_{k=0}^{\frac{n}{4}-1} \wedge^{2k+1} \varphi \quad \text{if } n \equiv 0 \pmod 4, \delta = 1 \\
\beta(\varphi) &= \langle\!\langle -a \rangle\!\rangle \widetilde{\gamma}(\varphi) \quad \text{if } n \geq 4, n \text{ even}, \delta = 1.
\end{aligned}$$

The forms $\widetilde{\gamma}(\varphi)$ are multiples of Pfister forms which are for $n > 2$ (!) hyperbolic if φ is isotropic. Since for $a, b \in D(\varphi)$ the product ab is a norm of a quadratic

[a] Rost is referring to the invariant b_1 from Section 20, which Serre had communicated to him in a letter.

[b] In the following equations, δ denotes the determinant $d(\varphi)$.

extension over which φ is isotropic, one sees that $a\widetilde{\gamma}(\varphi) \cong b\widetilde{\gamma}(\varphi)$ and $\langle\langle -a\rangle\rangle\widetilde{\gamma}(\varphi) = \langle\langle -b\rangle\rangle\widetilde{\gamma}(\varphi)$.

Hence these considerations give new definitions of α, β, γ without chain-equivalence of diagonalizations.

In any case $\beta(\varphi)$ can be viewed as a subspace of $\wedge\varphi$ of half dimension which is related to the $*$-operator,[c] or with the self-duality $\wedge\colon \wedge V \otimes \wedge V \to \wedge^n V$. This can be made more precise in the case $n \equiv 0 \pmod 4$ as follows.

Identify $\wedge V$ with the Clifford algebra $C(\varphi)$ in the natural way by $v_1 \wedge \cdots \wedge v_k \mapsto v_1 \ldots v_k$ for pairwise orthogonal v_i. Then the $*$-operator on $\wedge V$ is given by the action of $Z = F \oplus \wedge^n V = C(\wedge^n \varphi)$ on $C(\varphi)$ say from the left. If $n \equiv 0 \pmod 4$ then $Z \simeq F[t]/(t^2 - \delta)$. Let

$$E = \{x \in \wedge V \otimes_F Z \mid (1 \otimes \omega)x = (\omega \otimes 1)x \text{ for } \omega \in \wedge^n V\}$$

be the "eigenspace of the star operator". Considering $\beta(\varphi)$ as an element of $W(F)/\langle\langle\delta\rangle\rangle \subset W(Z)$ one finds

$$\langle 2\rangle \cdot \beta(\varphi) = [\wedge\varphi \otimes_F Z \mid E] \in W(Z).$$

Hence for $n \not\equiv 2 \pmod 4$ there is a natural construction of $\beta(\varphi)$, which implies e.g. that β can also be defined for quadratic forms over schemes. I don't know whether this is possible for $n \equiv 2 \pmod 4$, $n > 2$.

II) Relation with trace forms of algebras

If $\sqrt{-1} \in F^*$ one may interpret α, β, γ also as trace forms of algebras: Let tr be the trace of the even Clifford algebra $C_0(\varphi)$ over its center and let $\pi\colon C(\varphi) = C_0(\varphi) \oplus C_1(\varphi) \to C_0(\varphi)$ be the projection. Then

(1) If n is odd, then $\beta(\varphi)$ is given by the trace form $C_0(\varphi) \to F$, $x \mapsto \operatorname{tr}(x^2)$.
(2) If n is even, then $\gamma(\varphi)$ is given by the trace form $C_0(\varphi) \to Z$, $x \mapsto \operatorname{tr}(x^2)$.
(3) If $n \equiv 0 \pmod 4$ put $\phi\colon C(\varphi) \to Z$, $x \mapsto \operatorname{tr}\pi(xx^t)$. Then ϕ and $\phi|_{C_0(\varphi)}$ are forms over Z representing $\beta(\varphi)$ and $\gamma(\varphi)$ respectively.

Similar remarks hold for $\alpha(\varphi)$.

Hence if $\sqrt{-1} \in F^*$ the invariants are closely related with the Hasse-Witt invariant. On the other extreme, if $F = \mathbf{R}$ then (the cohomological versions of) α, β, γ evaluated on the anisotropic part of a form give a complete set of invariants. Maybe the assumption $\sqrt{-1} \in F^*$ is not very good in considering α, β, γ.

III) Cohomological interpretation of the trace form

Suppose $\sqrt{-1} \in F^*$. Then $x^n = 0$ for $x \in K_2^M F/2$ and $n > 1$, but there are operations

$$T_n\colon K_2^M F/2 \to K_{2n}^M F/2$$

[c]The $*$-operator is defined as follows, cf. [KMRT 98, 10.22]: Let V be a k-vector space of dimension n endowed with a nondegenerate quadratic form q of discriminant 1. Fix a nonzero element e in $\wedge^n V$ such that $(\wedge^n q)(e) = 1$; the wedge product determines a nondegenerate pairing $\wedge^i V \times \wedge^{n-i} V \to k$. The form q gives a nondegenerate form $\wedge^i q$ on $\wedge^i V$, hence we have $\wedge^i V \cong (\wedge^i V)^*$. Combining these two facts gives an isomorphism $\wedge^i V \cong \wedge^{n-i} V$. Since $(\wedge^n q)(e) = 1$, this is an isometry of $\wedge V$; it is known as the $*$-operator and is of much use in Hodge theory.

When $n = 2m$ is even, the automorphism of $\wedge^m V$ that we get has square 1 if m is even, and square -1 if m is odd. In the first case (i.e., n divisible by 4) the eigenspaces of the $*$-automorphism give an orthogonal decomposition of $\wedge^{n/2} V$ as two subspaces of the same dimension; the corresponding quadratic forms are isomorphic to the form $\ell(q)$ of 27.19.

(which may be called "reduced n-fold power") such that

(A.1) $$T_n(\sum_i \{a_i, b_i\}) = \sum_{i_1 < \cdots < i_n} \{a_{i_1}, b_{i_1}, \ldots, a_{i_n}, b_{i_n}\}.$$

To define T_n take (A.1) as a definition of T_n on formal sums of symbols $\{a, b\} \in F^* \times F^*$ and check $T_n(x + \{a, bc\}) = T_n(x + \{a, b\} + \{a, c\})$ etc.

Now note that

$$T_n\Big(\sum_{i=1}^{n} \{a_i, b_i\}\Big) = \{a_1, b_1, \ldots, a_n, b_n\}$$

while the trace form of $A = (a_1, b_1) \otimes \cdots \otimes (a_n, b_n)$ is $\langle\langle a_1, b_2, \ldots, a_n, b_n \rangle\rangle$.

Hence the T_n are (cohomological) operations which generalize the trace form construction for products of quaternion algebras.

IV) Interpretation of $\beta(\varphi)$ for $\dim \varphi = 3, 4, 5$

I noticed the following:

LEMMA. (1) *If* $\dim \varphi = 3$ *and* $\delta(\varphi) = -1$, *then*

$$\beta(-\varphi) = 0 \Leftrightarrow 1 \in D(\varphi).$$

(2) *If* $\dim \varphi = 4$ *and* $\delta(\varphi) = 1$, *then*

$$\beta(-\varphi) = 0 \Leftrightarrow 1 \in D(\varphi).$$

(3) *If* $\dim \varphi = 5$ *and* $\delta(\varphi) = -1$, *then the following are equivalent*:

(a) $D(\varphi) \cap [D(\varphi) \cdot D(\varphi)] \neq \emptyset$.

(b) $\beta(-\varphi) = 0$.

(c) $D(\varphi) \subset D(\varphi)D(\varphi)$.

Clearly a normalization of $\delta(\varphi)$ for odd $\dim \varphi$ is no restriction at all.

PROOF OF (3): Let $a \in D(\varphi)$. Then

$$\varphi \simeq \langle a_1, \ldots, a_5 \rangle$$

with $a_1 = a$. Moreover,

$$\beta(-\varphi) = \langle 1, -a_i, a_i a_j \rangle; \quad 1 \le i < j \le 5$$

and

$$\psi = a\varphi \perp -\varphi$$

is a subform of $\beta(-\varphi)$ of dimension $10 > \frac{1}{2} \dim \beta(-\varphi)$. Hence $\beta(-\varphi) = 0 \Leftrightarrow \psi$ is isotropic $\Leftrightarrow a \in D(\varphi)D(\varphi)$. □

[. . .]

M. Rost

APPENDIX B

A letter from J-P. Serre to R.S. Garibaldi

Paris, September 15, 2002

Dear Skip,

Here are a few historical comments on my involvement with cohomological invariants and trace forms.

Let me begin with the trace form.

I started being interested in it because of a letter from J.D.Gray, see [Se 78]. He wanted a modernized (?) account of Klein's reduction of the general quintic equation to the special form $x^5 + ax^2 + bx + c = 0$, together with its connections to modular functions of level 5. I noticed that this is possible thanks to a formula relating the two kinds of Stiefel-Whitney classes w_2 attached to an étale algebra E, namely the Galois one (which we now call w_2^{gal}), and the Delzant one for the quadratic trace form q_E of E. I wrote this in the form of a letter to J. Martinet (February 1982). I then went to Barcelona (May 1982), where Nuria Vila, a student of Pilar Bayer, was trying to solve the lifting problem (from A_n to $2.A_n$) for specific equations: trinomial ones, mainly. This amounts to deciding whether the Stiefel-Whitney class w_2^{gal} is 0. I explained to her the recipe of my letter to Martinet, which reduces the computation to that of $w_2(q_E)$, which is easy. This encouraged me to publish the formula (in [Se 84]), and people immediately started using it (see the references in 25.14). In [Se 84], I had left open the problem of generalizing the formula to an arbitrary w_i; this was soon solved by B. Kahn [Kahn 84]. A few years went by. Then I collaborated with E. Bayer on "self-dual normal bases", a closely related topic. This took place around 1991–92, but was published three years later [BS 94], because of the heavy backlog of Amer. J. Math. At about the same time (1991–92) I also started being interested in the possible form of q_E in low rank. Rank up to 5 is easy, but it made me notice that $w_3(q_E)$ is always 0. To go further I introduced in a short note (sent in July 1992 to E. Bayer and a few other friends) the notion of cohomological invariant — with the present definition, except that Rost's compatibility theorem 11.1 was then an axiom. This allowed me to show that the w_i are 0 beyond roughly half the rank. I also did the same with what I dubbed the "Witt invariants", which are supposed to give more precise information than the Stiefel-Whitney ones. I also started at that time the determination of all q_E of rank ≤ 7, although the most precise results were only obtained around 1994. I gave several courses on this: Harvard (1990 and 1992), Ascona (1991), Collège de France (1992 and 1994), Besançon (1993), Edinburgh (1995), the latest being the present one (UCLA, 2001). The fact that I did not publish until now anything other than short résumés was mostly due to the lack of proper foundational material (found here as Chapters II through V).

Now, about cohomological invariants.

The notion seems reasonable enough. It brings some order to all the many known invariants. It explains for instance why nobody ever came up with a non-constant invariant (mod 3) for étale algebras. On the other hand, most of the cases treated in the present notes are basically trivial.[a] They are relative to cases where there is a versal object over a purely transcendental function field: hence any invariant is characterized by its residues. One then uses the fact that one already knows enough such invariants to fill in all possibilities for the residues. We rarely have to construct new invariants from scratch (a few exceptions: 20.1, 25.16, 27.19, 27.22). This is very different from Merkurjev's part, where the Rost invariant has to be constructed the hard way. Incidentally, I had also tried in 1992 to construct the same invariant by a similar method (making use of the étale cohomology of $\mathbf{SL}_n/G$); however, I was unable to show that "my" invariant was not identically 0 (even for the simple case of G_2, how sad ...). I corresponded with Rost on this in December 1992 and found that he had gone much further; not only did he know his invariant is not always 0, but he could define it as taking values in $\mathbf{Q}/\mathbf{Z}(2)$, and he could relate it to his space "$Q(G)$" of quadratic forms. It was particularly satisfying to see a new invariant taking values in H^3, instead of the stale H^1 and H^2 of the previous decades. As Eva Bayer explained to me on several occasions, the ground had been well-prepared by the deep work of Pfister, Merkurjev-Suslin, and others, who definitely went beyond H^2. I also benefited from the work of (and my correspondence with) Petersson-Racine on F_4.

With best wishes — and many thanks for your enormous help (without which these notes would never have been written),

Yours,

J-P.Serre

[a]The main exception is the one I treated first, namely that of étale algebras. Here we use a subfunctor of Et_n (namely the multiquadratic algebras) which is strong enough to detect all the invariants, without being equal to Et_n itself.

APPENDIX C

A letter from B. Totaro to J-P. Serre

11 December 2002

Dear Serre,

Here is a geometric description of the group of cohomological invariants. To me, it seems like a natural companion to the Compatibility Theorem 11.1 in your notes.

LEMMA. *Let G be a smooth algebraic group over a field k_0, and let C be a finite Γ_{k_0}-module of order invertible in k_0. Let V be a representation of G over k_0, and suppose that V has a G-invariant Zariski-open subset U, with $V - U$ of codimension ≥ 2,*[a] *such that U is a G-torsor. Then the group* $\mathrm{Inv}(G, H^i(C))$ *of cohomological invariants is isomorphic to the subgroup of $H^i(k_0(U/G), C)$ consisting of the elements which are unramified along each irreducible divisor in U/G.*

PROOF. Your notes show that $\mathrm{Inv}(G, H^i(C))$ is a subgroup of $H^i(k_0(U/G), C)$ (since the G-torsor over the function field $k_0(U/G)$ is versal, 5.3, together with 12.3), and that every element of this subgroup is unramified along each irreducible divisor in U/G (by the Compatibility Theorem 11.1).

It remains to show that an element x of $H^i(k_0(U/G), C)$ which is unramified along each divisor in U/G belongs to the subgroup $\mathrm{Inv}(G, H^i(C))$. So, given any G-torsor E over an extension field k of k_0, we need to construct an element of $H^i(k, C)$ associated with x. First, "extending scalars", x determines an element of $H^i(k(U_k/G_k), C)$, which I will also call x. It is unramified along each irreducible divisor in U_k/G_k. (To check this, one can use that the morphism $U_k/G_k \to U/G$ is flat, so that every divisor in U_k/G_k either dominates U/G or dominates some divisor in U/G.)

Consider the diagram of k-varieties:

$$\operatorname{Spec} k = E/G_k \leftarrow (E \times U_k)/G_k \to U_k/G_k.$$

We can pull x back from the cohomology of the function field of U_k/G_k to the cohomology of the function field of $(E \times U_k)/G_k$. The resulting element is unramified along each irreducible divisor in $(E \times U_k)/G_k$. (Again, this follows from the fact that the morphism $(E \times U_k)/G_k \to U_k/G_k$ is flat.)

Since $V - U$ has codimension ≥ 2 in V, we have an element of the cohomology of the function field of $(E \times V_k)/G_k$ which is unramified along each irreducible divisor; the divisors here are the same as those in $(E \times U_k)/G_k$. Now $(E \times V_k)/G_k$ is a vector bundle over $E/G_k = \operatorname{Spec} k$. And your notes show (10.1) that an element of the cohomology of the function field of a vector space over k which is unramified

[a]This hypothesis is harmless; if a given representation V does not satisfy it, then one may replace V with $V \times V$.

along each divisor belongs to the cohomology of k itself. Thus we have produced an element of $H^i(k, C)$ associated with an element x as above and a G-torsor over k.

Thus, every unramified element x as above gives an element of $\mathrm{Inv}(G, H^i(C))$; the necessary functoriality property as k varies is automatic from the above construction. To complete the proof, we will show that applying this cohomological invariant to the natural G-torsor over the field $k_0(U/G)$ gives the element x in $H^i(k_0(U/G), C)$. By going through the construction, we see that it suffices to show that the two projection maps from $(U \times U)/G$ to U/G induce the same pullback homomorphism from unramified cohomology classes on U/G (meaning elements of $H^i(k_0(U/G), C)$ which are unramified on each divisor in U/G) to unramified cohomology classes on $(U \times U)/G$.

To prove this, we can argue that the two projection maps from $(U \times U)/G$ to U/G are homotopic in a suitable sense. To be precise, let

$$W = \{(t, x, y) \in \mathrm{Aff}^1 \times (U \times U)/G : tx + (1-t)y \in U\}.$$

Then W is an open subset of $\mathrm{Aff}^1 \times (U \times U)/G$ whose complement has codimension at least 2. We have a flat morphism $W \to U/G$ defined by

$$(t, x, y) \mapsto tx + (1-t)y.$$

So we have a pullback homomorphism from the unramified cohomology of U/G to that of W, or equivalently of $\mathrm{Aff}^1 \times (U \times U)/G$. By 10.1 in your notes, every unramified cohomology class on $\mathrm{Aff}^1 \times (U \times U)/G$ is "constant", that is, pulled back from a class on $(U \times U)/G$. In particular, the pullback to W of any unramified cohomology class on U/G has the same restriction to $t = 0$ as to $t = 1$. This means that the pullback homomorphisms on unramified cohomology associated to the two projection maps from $(U \times U)/G$ to U/G are equal. □

B. Totaro

Rost invariants of simply connected algebraic groups

By Alexander Merkurjev

With a section by Skip Garibaldi

Contents

Introduction

In the first note in this volume, the cohomological invariants (with, say, mod 2 coefficients) were determined for $\mathbf{O}_n$, $\mathbf{SO}_n$, S_n, ... Here we will describe the cohomological invariants (of degree 3 with coefficients in $\mathbb{Q}/\mathbb{Z}(2)$) of an arbitrary semisimple simply connected group. This description is due to M. Rost, cf. p. 98.

The contents are as follows. In Section 1 the definition of an invariant of an algebraic group G with values in a functor H is given. In Section 2 we consider cohomological invariants. In the following section we prove that under a mild restriction on H every invariant of G with values in H is uniquely determined by its value at the generic principal homogeneous space of G.

In Section 4 we review basic facts about the cycle modules defined by M. Rost in [Rost 96]. In the following two sections we compute certain K-cohomology groups of algebraic tori, simply connected groups, and their principal homogeneous spaces. The group of primary interest is $A^1(G, K_2)$ for a simply connected group G. This group was originally computed by P. Deligne in [Deli 79] (see also [EKLV 98]). It is isomorphic to the group $Q(G)$ of quadratic functions on the set of loops in G, discussed in Section 7.

In Section 8 we prove a theorem of B. Kahn which builds a bridge between K-cohomology (the group $A^1(G, K_2)$) and dimension 3 Galois cohomology with coefficients in $\mathbb{Q}/\mathbb{Z}(2)$. The Rost invariant of an absolutely simple simply connected group is defined in the following section. In Section 10 we compute the order n_G of the Rost invariant of a group G in terms of the representation theory of G.

In Section 11 through 15 the values of the numbers n_G are given for all (absolutely simple simply connected) groups of classical type. In the last section we compute the numbers n_G for groups of exceptional type. This section was written by S. Garibaldi. The definition and simple properties of the groups $H^{d+1}(F, \mathbb{Q}/\mathbb{Z}(d))$ for a field F are given in Appendix A.

The letter F is reserved for the base field. We don't impose any restriction on the characteristic of F. We write $\mathrm{Gal}(F)$ for the absolute Galois group $\mathrm{Gal}(F_{\mathrm{sep}}/F)$ of a separable closure F_{sep} over F. If A is a central simple algebra over a field F, we let $\exp(A)$ and $\mathrm{ind}(A)$ be the *exponent* and *index* of A respectively. We write $M\{p\}$ for the p-primary component of a torsion abelian group M. If X is an algebraic variety (a reduced separated scheme of finite type) over a field F and E/F is a field extension, we write X_E for the variety $X \times_{\mathrm{Spec}\, F} \mathrm{Spec}\, E$ over E and $X(E)$ for the set $\mathrm{Mor}_F(\mathrm{Spec}\, E, X)$ of E-points of X. We let X_{sep} denote $X_{F_{\mathrm{sep}}}$ for a separable closure F_{sep} of F. An *algebraic group* over F is a smooth affine group scheme of finite type over F.

1. Definition of an invariant

Let G be an algebraic group defined over a field F. If E/F is a field extension, we write $H^1(E, G)$ for the cohomology set $H^1(\mathrm{Gal}(E), G(E_{\mathrm{sep}}))$ (see [Se 65] and [KMRT 98, Ch. VII]). We identify $H^1(E, G)$ with the set of isomorphism classes of principal homogeneous spaces of G_E over E.

Consider the functor

$$H^1(*, G) : \mathbf{Fields}_{/F} \to \mathbf{Sets}, \quad E/F \mapsto H^1(E, G)$$

from the category of field extensions over F to the category of sets.

Let

$$H : \textbf{Fields}_{/F} \to \textbf{Abelian Groups}$$

be another functor to the category of abelian groups. An *invariant of G with values in H* (or an *H-invariant of G*) is a morphism of functors

$$a : H^1(*, G) \to H,$$

or, equivalently, a collection of maps $a_E : H^1(E, G) \to H(E)$ for every field extension E/F such that for every field homomorphism $i : E \to E'$ over F the diagram

$$\begin{array}{ccc} H^1(E,G) & \xrightarrow{a_E} & H(E) \\ {\scriptstyle H^1(i,G)}\downarrow & & \downarrow{\scriptstyle H(i)} \\ H^1(E',G) & \xrightarrow{a_{E'}} & H(E') \end{array}$$

is commutative.

The set $\mathrm{Inv}(G, H)$ of all H-invariants of G has the structure of an abelian group. An algebraic group homomorphism $\varphi : G \to G'$ induces a homomorphism

$$\varphi^* : \mathrm{Inv}(G', H) \to \mathrm{Inv}(G, H).$$

Thus, $\mathrm{Inv}(*, H)$ is a contravariant functor from the category of algebraic groups over F to **Abelian Groups**.

An element $h \in H(F)$ defines a *constant* H-invariant a^h of G as follows:

$$a^h_E(x) = H(i)(h) \quad \text{for every } x \in H^1(E, G),$$

where $i : F \to E$ is the inclusion. The subgroup of constant invariants $\mathrm{Inv}(G, H)_{\mathrm{const}}$ in $\mathrm{Inv}(G, H)$ is isomorphic to $H(F)$.

An invariant with values in H is called *normalized* if it takes the distinguished element of $H^1(F, G)$ to zero. We denote the subgroup of normalized invariants by $\mathrm{Inv}(G, H)_{\mathrm{norm}}$. Clearly,

$$\mathrm{Inv}(G, H) = \mathrm{Inv}(G, H)_{\mathrm{const}} \oplus \mathrm{Inv}(G, H)_{\mathrm{norm}}.$$

Note that the functors $H^1(*, G)$ and H extend naturally to the category of étale F-algebras as well as every invariant with values in H.

2. Invariants with values in $H^{d+1}\big(*, \mathbb{Q}/\mathbb{Z}(d)\big)$

Let F be a field. For every $d \geq 0$ we consider the functor H^{d+1} defined by

$$H^{d+1}(E) = H^{d+1}\big(E, \mathbb{Q}/\mathbb{Z}(d)\big)$$

for a field extension E/F (see App. A for a definition). We write $\mathrm{Inv}^{d+1}\big(G, \mathbb{Q}/\mathbb{Z}(d)\big)$ for the group of invariants $\mathrm{Inv}(G, H^{d+1})$, where G is an algebraic group over F.

EXAMPLE 2.1. Let G be a semisimple algebraic group over F and let C be the kernel of the universal cover $\widetilde{G} \to G$. For a character $\rho \in C^* = \mathrm{Hom}(C, \mathbf{G}_{\mathrm{m}})$, consider the commutative diagram

$$\begin{array}{ccccccccc} 1 & \longrightarrow & C & \longrightarrow & \widetilde{G} & \longrightarrow & G & \longrightarrow & 1 \\ & & {\scriptstyle \rho}\downarrow & & \downarrow & & \| & & \\ 1 & \longrightarrow & \mathbf{G}_{\mathrm{m}} & \longrightarrow & G_1 & \longrightarrow & G & \longrightarrow & 1 \end{array}$$

with exact rows and where the left square is a pushout. The connecting map in the cohomology sequence associated to the bottom short exact sequence gives rise to an invariant in $\operatorname{Inv}^2(G, \mathbb{Q}/\mathbb{Z}(1))$. Thus, we get a map (group homomorphism)

$$C^* \to \operatorname{Inv}^2(G, \mathbb{Q}/\mathbb{Z}(1)).$$

EXAMPLE 2.2. Let A be a central simple algebra over F. We write $\mathbf{SL}_1(A)$ for the kernel of the reduced norm character $\operatorname{Nrd} : \mathbf{GL}_1(A) \to \mathbf{G}_{\mathrm{m}}$. The connecting map in the cohomology sequence associated to the short exact sequence

$$1 \to \mathbf{SL}_1(A) \to \mathbf{GL}_1(A) \xrightarrow{\operatorname{Nrd}} \mathbf{G}_{\mathrm{m}} \to 1$$

and Hilbert's Theorem 90 yield a canonical bijection of pointed sets

$$H^1(F, \mathbf{SL}_1(A)) \xrightarrow{\sim} F^\times / \operatorname{Nrd}(A^\times).$$

We can define a normalized invariant $r \in \operatorname{Inv}^3(\mathbf{SL}_1(A), \mathbb{Q}/\mathbb{Z}(2))$ by the formula

$$r(a \operatorname{Nrd}(A^\times)) = \{a\} \cup [A] \in H^3(F, \mathbb{Q}/\mathbb{Z}(2)),$$

where $[A]$ is the class of the algebra A in the Brauer group $\operatorname{Br}(F) = H^2(F, \mathbb{Q}/\mathbb{Z}(1))$ (for the latter identification and definition of the cup product, see Appendix A). Note that $\{a\} \cup [A]$ does not depend on the choice of representative a: Each $n \in \operatorname{Nrd}(A^\times)$ is a norm $N_{L/F}(\ell)$ for some field L such that $A \otimes L$ is split and some $\ell \in L^*$. Then

$$\{n\} \cup [A] = \{N_{L/F}(\ell)\} \cup [A] = \operatorname{Cor}_{L/F}(\{\ell\} \cup [A \otimes L])$$

by the projection formula. Since $[A \otimes L] = 0$, we have that r is well-defined.

EXAMPLE 2.3 (Arason invariant). For an element $a \in F^\times$ we write $\langle\langle a \rangle\rangle$ for the binary bilinear form with diagonal matrix $\operatorname{diag}(1, -a)$. If E is an étale quadratic F-algebra, we let n_E be the (binary) norm quadratic form on E over F. For a_1, a_2, …, $a_{n-1} \in F^\times$ and an étale quadratic F-algebra E let

$$g = \langle\langle a_1, a_2, \dots, a_{n-1}; E \rangle\rangle$$

be the tensor product of the binary forms $\langle\langle a_1 \rangle\rangle$, …, $\langle\langle a_{n-1} \rangle\rangle$ and the quadratic form n_E. The quadratic form g is called an *n-fold Pfister form.*

A quadratic form q is *nondegenerate* if the radical $V^\perp$ of the associated symmetric bilinear form is 0 or if $\dim V^\perp = 1$ and $q(V^\perp) \neq 0$. (The latter possibility occurs only if $\operatorname{char} F = 2$ and $\dim V$ is odd.) Let $W_q(F)$ be the Witt group of classes of nondegenerate quadratic forms over F (see [Lam 73] and [Bae 01]). For every $n \geq 1$, let $I^n(F)$ denote the subgroup of $W_q(F)$ generated by the classes of all n-fold Pfister forms.[a]

For a Pfister form $g = \langle\langle a_1, a_2, \dots, a_{n-1}; E \rangle\rangle$, we set

$$e_n(g) = \{a_1, a_2, \dots, a_{n-1}\} \cup [E] \quad \text{in } H^n(F, \mathbb{Q}/\mathbb{Z}(n-1)),$$

where $[E]$ is the class of E in $H^1(F, \mathbb{Q}/\mathbb{Z}) = \operatorname{Hom}_{\mathrm{cont}}(\operatorname{Gal}(F), \mathbb{Q}/\mathbb{Z})$.

The class e_3 of a 3-fold Pfister form extends to a well-defined group homomorphism (see [Ar 75] if $\operatorname{char} F \neq 2$ and [Kato 82] if $\operatorname{char} F = 2$)

$$e_3 : I^3(F) \to H^3(F, \mathbb{Q}/\mathbb{Z}(2)).$$

Let q be a nondegenerate quadratic form of dimension n over F. We write $\mathbf{Spin}(q)$ for the spinor group and $\mathbf{O}(q)$ for the orthogonal group of q. There is a natural

[a]The papers [Bae 01], [Kato 82], and [Sah 72] write $I^{n-1}W_q(F)$ instead of $I^n(F)$.

homomorphism $\mathbf{Spin}(q) \to \mathbf{O}(q)$ [KMRT 98, p. 349]. It induces a map of pointed sets

$$f : H^1\big(F, \mathbf{Spin}(q)\big) \to H^1\big(F, \mathbf{O}(q)\big).$$

The latter set is in canonical bijection with the set $\mathrm{Quad}_n(F)$ of isomorphism classes of nondegenerate quadratic forms of dimension n over F. The image of f is identified then with the subset of $\mathrm{Quad}_n(F)$ consisting of classes of quadratic forms with the same discriminant (Arf invariant if $\operatorname{char} F = 2$) and the same Clifford invariant as q [KMRT 98, §31].

Let $u \in H^1\big(F, \mathbf{Spin}(q)\big)$. The element $f(u) \in H^1\big(F, \mathbf{O}(q)\big)$ corresponds to the isomorphism class of a quadratic form q' such that q and q' have the same discriminant and Clifford invariant. Then the form $h = q' \perp (-q)$ has trivial discriminant and Clifford invariant and therefore, $[h]$ lies in $I^3(F)$ (see [Mer 81] if $\operatorname{char} F \neq 2$ and [Sah 72, Th. 2(ii)] if $\operatorname{char} F = 2$). The cohomology class $e_3(h) \in H^3\big(F, \mathbb{Q}/\mathbb{Z}(2)\big)$ is well-defined and the correspondence $u \mapsto e_3(h)$ gives rise to the *Arason invariant* in $\mathrm{Inv}^3\big(\mathbf{Spin}(q), \mathbb{Q}/\mathbb{Z}(2)\big)$.

3. Classifying varieties of algebraic groups

A (connected) algebraic group S over F is called *special* if $H^1(E, S) = 1$ for every field extension E/F.

EXAMPLE 3.1. The groups $\mathbf{GL}_n$, $\mathbf{SL}_n$, $\mathbf{Sp}_{2n}$ and their products are special.

Clearly, $\mathrm{Inv}(S, H)_{\mathrm{norm}} = 0$ for every H if S is special.

Let G be an algebraic group over F. Choose an injective group homomorphism ρ of G into a special group S and consider the factor variety $X = S/\rho(G)$. Since S is special, for every field extension E/F, the set $H^1(E, G)$ can be identified with the orbit space of the action of $S(E)$ on $X(E)$ [KMRT 98, Cor. 28.4]:

$$H^1(E, G) = X(E)/S(E).$$

The variety X is called a *classifying variety of* G. The $S(E)$-orbits in the set $X(E)$ classify principal homogeneous spaces of G over E.

We write Y_p for the principal homogeneous space corresponding to a point $p \in X(E)$. If x is a point of X, Y_x denotes the corresponding principal homogeneous space over the residue field $F(x)$.

The principal homogeneous space Y_ξ for the generic point $\xi \in X$ is the generic fiber of the universal G-torsor $S \to X$ over the function field $F(X)$. We call Y_ξ a *generic principal homogeneous space of* G.

Let

$$H\colon \mathbf{Fields}_{/F} \to \mathbf{Abelian\ Groups}$$

be a functor. We have a homomorphism

$$\theta_{G,H} : \mathrm{Inv}(G, H) \to H\big(F(X)\big), \quad a \mapsto a_{F(X)}(Y_\xi).$$

We would like to prove that the map $\theta_{G,H}$ is injective (i.e., that an invariant of G is defined by its value at the generic principal homogeneous space Y_ξ) under a mild restriction on H, namely,

(∗) *For every field extension E/F, the homomorphism $H(E) \to H\big(E((t))\big)$ is injective.*

Note that by Proposition A.9 the functor $E \mapsto H^{d+1}\big(E, \mathbb{Q}/\mathbb{Z}(d)\big)$ satisfies this condition.

LEMMA 3.2. *Let x and x' be two points of X such that x' is regular of codimension 1 in the closure of x and the field extension $F(x')/F$ is separable. Suppose that for an invariant $a \in \mathrm{Inv}(G, H)$ one has $a_{F(x)}(Y_x) = 0 \in H\big(F(x)\big)$. If the functor H satisfies $(*)$, then $a_{F(x')}(Y_{x'}) = 0 \in H\big(F(x')\big)$.*

PROOF. Write A for the local ring of the point x' in the closure of x. By assumption, A is a discrete valuation ring with quotient field $F(x)$ and residue field $F(x')$; write $\widehat{A}$ for its completion. Since $F(x')$ is separable over F, it can be lifted to $\widehat{A}$ as in the proof of 11.6 on p. 30. Thus $\widehat{A}$ is isomorphic to $F(x')[[t]]$ over F; its quotient field E is isomorphic to $F(x')((t))$. We have the diagram of obvious ring homomorphisms

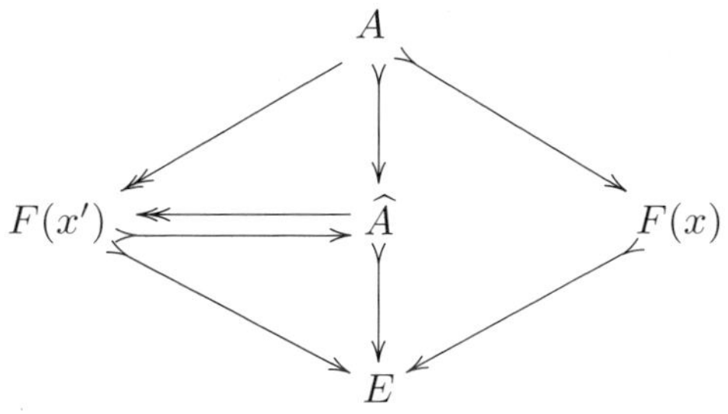

with three commutative triangles. It induces the diagram of maps

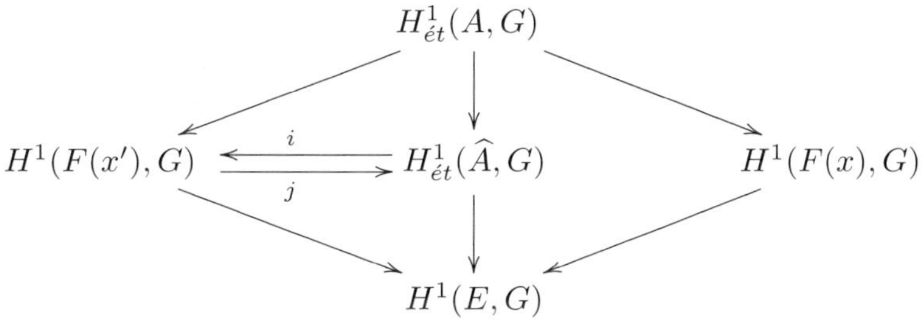

All three triangles in the diagram commute. By [SGA 3, Exp. XXIV, Prop. 8.1], the maps i and j are two bijections inverse to each other. Hence, the square is also commutative.

The element of the set $H^1_{ét}(A, G)$ representing the pullback of the universal torsor $S \to X$ with respect to the canonical morphism $\mathrm{Spec}\, A \to X$ has images Y_x and $Y_{x'}$ in $H^1\big(F(x), G\big)$ and $H^1\big(F(x'), G\big)$ respectively. Let Y be the image of the universal torsor in $H^1(E, G)$. By assumption, $a_{F(x)}(Y_x) = 0$, hence $a_E(Y) = 0 \in H(E)$. Therefore, $a_{F(x')}(Y_{x'})$ belongs to the kernel of $H\big(F(x')\big) \to H(E)$, which is trivial since H satisfies $(*)$. $\square$

THEOREM 3.3. *If H satisfies $(*)$, the map $\theta_{G,H}$ is injective.*

PROOF. Assume that for an invariant $a \in \mathrm{Inv}(G, H)$ we have $a_{F(X)}(Y_\xi) = 0$, where ξ is the generic point of X. Choose a field extension E/F and any $p \in X(E)$, i.e., a morphism $p : \mathrm{Spec}(E) \to X$. We have to show that $a_E(Y_p) = 0$. Denote by $x \in X$ the only point in the image of p. Replacing the base field F by E and the variety X by X_E we may assume that x is a rational closed point.

Let O be the (smooth) local ring of x in X and let $u_1, u_2, \dots, u_n$ be a system of local parameters of O. For $i = 0, 1, \dots, n$ write x_i for the generic point of the subvariety of X defined by the equations $u_1 = \dots = u_i = 0$ in a neighborhood of x. We have $x_0 = \xi$, $x_n = x$, and x_{i+1} is regular of codimension 1 in the closure of x_i for all $i = 0, 1, \dots, n-1$. The residue field $F(x_i)$ is the quotient field of the smooth local ring $O/(u_1, \dots, u_i)$ and hence is a separable extension of F. By Lemma 3.2 (applied n times), $a_{F(x)}(Y_x) = 0$, in other words, $a_E(Y_p) = 0$. □

4. Cycle modules and cohomology

We review the basic facts about cycle modules defined by M. Rost in [Rost 96]. Let X be an algebraic variety over F. A *cycle module* over X is a collection of $\mathbb{Z}$-graded abelian groups $M(x)_*$ for every point $x \in X(E)$ for all field extensions E/F with some additional data satisfying certain rules. The data consists of:

(1) Degree zero restriction homomorphisms $M(x)_* \to M(y)_*$ for every pair of points $x \in X(L)$ and $y \in X(E)$ and a field homomorphism $f : L \to E$ over F such that $X(f)(x) = y$;
(2) Degree zero corestriction homomorphisms $M(y)_* \to M(x)_*$ for every pair of points $x \in X(L)$ and $y \in X(E)$ and a finite field homomorphism $f : L \to E$ over F such that $X(f)(x) = y$;
(3) A left $K_*(E)$-module structure on $M(x)_*$ for every $x \in X(E)$;
(4) A degree -1 *residue homomorphism* with respect to discrete valuations.

If $Y \to X$ is a morphism of varieties over F, then a cycle module over X can be restricted to a cycle module over Y.

A cycle module M over $\operatorname{Spec} F$ can be viewed as a collection of functors

$$M_i\colon \mathbf{Fields}_{/F} \to \mathbf{Abelian\ Groups}, \quad E \mapsto M_i(E),$$

so that we can consider invariants of algebraic groups with values in cycle modules.

EXAMPLE 4.1. Milnor's K-groups $K_n(E)$ for all field extensions E/F form a cycle module over $\operatorname{Spec} F$.

EXAMPLE 4.2. If p is a prime integer different from $\operatorname{char} F$, then the assignment

$$E \mapsto H^{d+1}\big(E, \mathbb{Q}_p/\mathbb{Z}_p(d)\big)$$

defines a *cohomological cycle module* over $\operatorname{Spec} F$ [Rost 96, Rem. 1.11].

Note that for a field F of characteristic $p > 0$ the functor $H^{d+1}\big(*, \mathbb{Q}_p/\mathbb{Z}_p(d)\big)$ is not a cycle module in general since the residue homomorphisms are not defined (see Appendix A).

Let X be an equidimensional algebraic variety over F, and let M_* be a cycle module over X. We write $X^{(i)}$ for the set of points of X of codimension i. For every p, the homology groups of the complex $C^*(X; M)$ [Rost 96, Sec. 5]

$$\to \coprod_{x \in X^{(p-1)}} M_{n-p+1}\big(F(x)\big) \xrightarrow{\partial} \coprod_{x \in X^{(p)}} M_{n-p}\big(F(x)\big) \xrightarrow{\partial} \coprod_{x \in X^{(p+1)}} M_{n-p-1}\big(F(x)\big) \to$$

are denoted by $A^p(X, M_n)$.

EXAMPLE 4.3. (1) The group $A^p(X, K_p)$ is the Chow group $\mathrm{CH}^p(X)$ of classes of codimension p cycles on X.

(2) If X is smooth, the bigraded group $A^*(X, K_*)$ has a natural ring structure [Rost 96, §14].

(3) If X is a normal variety, the group $A^0(X, K_1)$ coincides with the group $F[X]^\times$ of invertible regular functions on X.

EXAMPLE 4.4. Let $\pi : Y \to X$ be a flat morphism and let M_* be a cycle module over Y. For every point $x : \operatorname{Spec} E \to X$, write $\pi^{-1}x$ for the fiber of π over x. For every $q \geq 0$, the assignment

$$x \mapsto A^q(\pi^{-1}x, M_*)$$

is a cycle module over X [Rost 96, Sec. 7].

5. K-cohomology of split tori and their principal homogeneous spaces

Let G be a connected algebraic group defined over a field F. The character group $G^* = \operatorname{Hom}_F(G, \mathbf{G}_\mathrm{m})$ of all characters $G \to \mathbf{G}_\mathrm{m}$ defined over F can be viewed as a subgroup of the group of invertible regular functions $F[G]^\times$. By [Rose 61, Th. 3],

$$F[G]^\times = F^\times \oplus G^*. \tag{5.1}$$

Hence, every invertible regular function f on G such that $f(1) = 1$ is a character of G, i.e., $f \in G^*$.

Let Y be a principal homogeneous space of G over F. Assume first that Y is trivial, i.e., Y has a point over F. Choose a point $y \in Y(F)$ and a function $h \in F[Y]^\times$ and consider the function f on G defined by

$$f(g) = h(gy) \cdot h(y)^{-1}$$

for every g in G. Clearly, $f \in F[G]^\times$ and $f(1) = 1$, hence $f \in G^*$. The character f does not depend on the choice of y. For if y' is another point in $Y(F)$, $y' = g'y$ for some $g' \in G(F)$ and

$$\begin{aligned} h(gy') \cdot h(y')^{-1} &= h(gg'y) \cdot h(g'y)^{-1} \\ &= h(gg'y) \cdot h(y)^{-1} \cdot h(y) \cdot h(g'y)^{-1} \\ &= f(gg') \cdot f(g')^{-1} \\ &= f(g). \end{aligned}$$

Thus, we get a well defined homomorphism

$$\kappa : F[Y]^\times \to G^*, \quad h \mapsto f.$$

By (5.1), we have a split exact sequence

$$1 \to F^\times \to F[Y]^\times \xrightarrow{\kappa} G^* \to 0. \tag{5.2}$$

Let Y be a (not necessarily trivial) principal homogeneous space of G over F. Since Y_sep is trivial, we have a homomorphism

$$\kappa_\mathrm{sep} : F_\mathrm{sep}[Y_\mathrm{sep}]^\times \to G^*_\mathrm{sep}.$$

Clearly, κ_sep is $\operatorname{Gal}(F)$-equivariant. Taking the Galois invariant elements, we get a homomorphism $\kappa : F[Y]^\times \to G^*$ defined for every Y. Hilbert's Theorem 90 then yields:

PROPOSITION 5.3. *For every principal homogeneous space Y of G over F the sequence* (5.2) *is exact.*

COROLLARY 5.4. *If G is a semisimple group, then $F[Y]^\times = F^\times$ for every principal homogeneous space Y of G.*

Let X be an equidimensional smooth variety over F. The bigraded group $A^*(X \times \mathbf{G}_{\mathrm{m}}, K_*)$ is a bimodule over $A^*(X, K_*)$. The projection $f : X \times \mathbf{G}_{\mathrm{m}} \to \mathbf{G}_{\mathrm{m}}$ can be viewed as an invertible regular function on $X \times \mathbf{G}_{\mathrm{m}}$ and hence can be considered as an element of $A^0(X \times \mathbf{G}_{\mathrm{m}}, K_1) = F[X \times \mathbf{G}_{\mathrm{m}}]^\times$.

PROPOSITION 5.5. *The (right) $A^*(X, K_*)$-module $A^*(X \times \mathbf{G}_{\mathrm{m}}, K_*)$ is free with basis $\{1, f\}$.*

PROOF. Let $i \colon X \to X \times \mathbf{A}^1_F$ be the closed embedding given by $i(x) = (x, 0)$, and let $j \colon X \times \mathbf{G}_{\mathrm{m}} \to X \times \mathbf{A}^1_F$ be the open embedding. The connecting homomorphism ∂ in the exact localization sequence [Rost 96, §5]

$$\cdots \xrightarrow{i_*} A^*(X \times \mathbf{A}^1_F, K_*) \xrightarrow{j^*} A^*(X \times \mathbf{G}_{\mathrm{m}}, K_*) \xrightarrow{\partial} A^*(X, K_{*-1}) \xrightarrow{i_*} \cdots$$

splits by the left multiplication by f [Rost 96, Rule R3d], hence $i_* = 0$. By the homotopy invariance property [Rost 96, Prop. 8.6], the homomorphism $p^* : A^*(X, K_*) \to A^*(X \times \mathbf{A}^1_F, K_*)$ induced by the projection $p : X \times \mathbf{A}^1_F \to X$ is an isomorphism. Therefore, the image of j^* coincides with $1 \cdot A^*(X, K_*)$. □

Let T be a split torus over F. Since $A^*(\operatorname{Spec} F, K_*) = K_*(F)$, Proposition 5.5 yields by induction:

COROLLARY 5.6. *Let $f_1, f_2, \dots, f_m$ be a basis of T^*. Then*

(1) *$A^0(T, K_*)$ is a free $K_*(F)$-module with basis consisting of the elements $\{f_{i_1}, f_{i_2}, \dots, f_{i_q}\}$ for $q = 0, 1, \dots, m$ and all q-tuples of indices $i_1 < i_2 < \dots < i_q$.*

(2) *$A^p(T, K_*) = 0$ if $p > 0$.*

5.7. Let S be a principal homogeneous space of the split torus T. By Hilbert's Theorem 90, S is trivial, i.e., S is isomorphic to T. We would like to compute the K-cohomology groups of S. By Corollary 5.6, $A^p(S, K_*) = 0$ if $p > 0$. Consider the filtration on $M_n = A^0(S, K_n)$:

$$0 = M_n^{(1)} \subset M_n^{(0)} \subset \cdots \subset M_n^{(-n+1)} \subset M_n^{(-n)} = M_n,$$

where $M_n^{(p)}$ is the image of the natural product map

$$K_{n+p}(F) \otimes M_{-p} \to M_n.$$

We write $M_n^{(p/p+1)}$ for the factor group $M_n^{(p)}/M_n^{(p+1)}$.

By Proposition 5.3, we have an exact sequence

$$1 \to F^\times \to F[S]^\times \xrightarrow{\kappa} T^* \to 0.$$

For every $n \geq 0$ and $p \leq 0$ define homomorphisms

$$\varphi_{n,p} : K_{n+p}(F) \otimes \Lambda^{-p} T^* \to M_n^{(p/p+1)}$$

by the formula

(5.8) $$\varphi_{n,p}\big(\{a_1, a_2, \dots, a_{n+p}\} \otimes (f_1 \wedge f_2 \wedge \dots \wedge f_{-p})\big) = \\ = \{a_1, a_2, \dots, a_{n+p}, h_1, h_2, \dots, h_{-p}\} + M_n^{(p+1)},$$

where the $h_i \in F[S]^\times$ satisfy $\kappa(h_i) = f_i$. It follows from Proposition 5.3 that φ does not depend on the choice of the h_i. Corollary 5.6.1 then yields:

PROPOSITION 5.9. *The maps $\varphi_{n,p}$ are isomorphisms.*

6. K-cohomology of simply connected algebraic groups and their principal homogeneous spaces

In this section we compute the group $A^1(G, K_2)$ for a split simply connected group G.

PROPOSITION 6.1. *Let Y be a principal homogeneous space of a simply connected semisimple group. Then the Chow groups $\mathrm{CH}^i(Y)$ are trivial for $i = 1, 2$.*

PROOF. Let $K_0(Y)^{(i)}$ be the i-th term of the topological filtration on $K_0(Y)$ [Qu 73, §5]. Set

$$K_0(Y)^{(i/i+1)} = K_0(Y)^{(i)}/K_0(Y)^{(i+1)}.$$

There are canonical surjective homomorphisms

$$b_i : \mathrm{CH}^i(Y) \to K_0(Y)^{(i/i+1)}$$

defined by $b_i([Z]) = [\mathcal{O}_Z]$ for a prime cycle $Z \subset Y$ of codimension i. The Chern class c_i [Fu 84, 3.2] considered as a map from $K_0(Y)$ to $\mathrm{CH}^i(Y)$ vanishes on $K_0(Y)^{(i+1)}$ [Fu 84, Ex. 15.3.1] and hence defines a homomorphism

$$c_i : K_0(Y)^{(i/i+1)} \to \mathrm{CH}^i(Y).$$

The Riemann-Roch theorem implies that the composition $c_i \circ b_i$ is multiplication by $(-1)^{i-1}(i-1)!$ [Fu 84, Ex. 15.3.1]. It follows that b_i and c_i are isomorphisms for $i = 1, 2$. On the other hand, by [Pa 98], $K_0(Y) = \mathbb{Z} \cdot 1$, hence $K_0(Y)^{(1)} = 0$ and therefore $K_0(Y)^{(i/i+1)} = 0$ for $i \geq 1$. □

6.2. Let G be a split simply connected group over F and let T be a split maximal torus of G. Let $X = G/T$ be the variety of (right) cosets and let $\pi : G \to X$ be the canonical morphism. For every point $x \in X(E)$ over a field extension E/F, the coset $\pi^{-1}x \subset G$ is a trivial principal homogeneous space of T over E. Consider the spectral sequence associated to the morphism π [Rost 96, Sec. 8]:

$$E_1^{p,q} = \coprod_{x \in X^{(p)}} A^q(\pi^{-1}x, K_{n-p}) \Longrightarrow A^{p+q}(G, K_n).$$

By Proposition 5.6.2, $E_1^{p,q} = 0$ if $q > 0$. Hence, we have isomorphisms

$$\gamma_G : A^p(G, K_n) \xrightarrow{\sim} A^p(X, M_n),$$

where M is the cycle module over X defined by

$$M_n(x) = A^0(\pi^{-1}x, K_n)$$

for a point $x \in X(E)$ (see Example 4.4).

Consider the following filtration of M_n by cycle submodules. Let $M_n^{(p)}(E)$ be the image of the natural product map

$$K_{n+p}(E) \otimes M_{-p}(E) \to M_n(E).$$

We have the filtration (see 5.7)

(6.3) $$0 = M_n^{(1)} \subset M_n^{(0)} \subset \cdots \subset M_n^{(-n+1)} \subset M_n^{(-n)} = M_n.$$

By Proposition 5.9, the homomorphism of cycle modules

(6.4) $$\varphi_{n,p} : K_{n+p} \otimes \Lambda^{-p}\, T^* \to M_n^{(p/p+1)}$$

defined as in (5.8), is an isomorphism. Hence, the E_1-term of the standard spectral sequence

$$E_1^{p,q} = A^{p+q}(X, M_n^{(p/p+1)}) \Longrightarrow A^{p+q}(X, M_n)$$

associated with the filtration on M is isomorphic to

$$A^{p+q}(X, K_{n+p}) \otimes \Lambda^{-p}\, T^*.$$

LEMMA 6.5. *The product homomorphism* $K_i(F) \otimes \mathrm{CH}^j(X) \to A^j(X, K_{i+j})$ *is an isomorphism.*

PROOF. Let $B \subset G$ be a Borel subgroup containing T. Every fiber the natural morphism $\alpha\colon X \to G/B$ is an affine space. Hence, the pullback homomorphisms $A^j(G/B, K_i) \to A^j(X, K_i)$ are isomorphisms by the homotopy invariance property. Thus, we only need to prove the statement when X is replaced by G/B. But G/B is a cellular variety and the result follows from [EKLV 98, Prop. 3.7]. □

By Lemma 6.5,

$$A^{p+q}(X, K_{n+p}) \cong K_{n-q}(F) \otimes \mathrm{CH}^{p+q}(X).$$

We have proved:

THEOREM 6.6. (Cf. [EKLV 98, 3.7]) *Let G be a split simply connected group, let $T \subset G$ be a split maximal torus, and set $X = G/T$. For every $n \geq 0$ there is a spectral sequence*

$$E_1^{p,q} = K_{n-q}(F) \otimes \mathrm{CH}^{p+q}(X) \otimes \Lambda^{-p}\, T^* \Longrightarrow A^{p+q}(G, K_n).$$

The spectral sequence is supported in the triangle defined by the inequalities $p \leq 0$, $p+q \geq 0$, and $n \geq q$.

Consider first the case $n = 1$. The nontrivial E_1-terms of the spectral sequence

$$\begin{array}{ccc} E_1^{-1,1} & \xrightarrow{\ d\ } & E_1^{0,1} \\ & & E_1^{0,0} \end{array}$$

coincide with

$$\begin{array}{ccc} T^* & \xrightarrow{\ d\ } & \mathrm{CH}^1(X) \\ & & F^\times \end{array}$$

As $A^1(G, K_1) = \mathrm{CH}^1(G)$ (Example 4.3.1), the spectral sequence yields an exact sequence

$$1 \to F^\times \to A^0(G, K_1) \to T^* \xrightarrow{d} \mathrm{CH}^1(X) \to \mathrm{CH}^1(G) \to 0.$$

Note that by (5.1) and Example 4.3.3,

$$A^0(G, K_1) = F[G]^\times = F^\times \oplus G^*.$$

Thus we have an exact sequence

$$0 \to G^* \to T^* \xrightarrow{d} \mathrm{CH}^1(X) \to \mathrm{CH}^1(G) \to 0,$$

where the homomorphism d is defined as follows: Let $Y_\xi = \pi^{-1}(\xi)$ be the generic fiber of π. It is a (trivial) principal homogeneous space of T over the field $F(X)$. By

Proposition 5.3, for a character $f \in T^*$ there is a function $h \in F(X)[Y_\xi]^\times \subset F(G)^\times$ such that $\kappa(h) = f$, i.e., such that

$$h(tg) = f(t)h(g) \quad \text{for all } t \text{ in } T \text{ and } g \text{ in } G.$$

The principal divisor $\operatorname{div}(h) \in \operatorname{Div}(G)$ is T-invariant, hence there is a unique divisor D on X such that $\pi^{-1}(D) = \operatorname{div}(h)$. Then $d(f) = [D]$. Note that h is uniquely determined up to a multiple in $F(X)^\times$, hence the class $[D] \in \mathrm{CH}^1(X)$ is well-defined.

If G is semisimple, then $G^* = 0$ and hence d is injective. If in addition G is simply connected, then $\mathrm{CH}^1(G) = 0$ (Proposition 6.1) and hence d is an isomorphism.

Now consider the case $n = 2$. We use the spectral sequence in Theorem 6.6 to compute the group $A^1(G, K_2)$. The nontrivial E_1-terms of the spectral sequence

$$\begin{array}{ccccc} E_1^{-2,2} & \xrightarrow{d_1} & E_1^{-1,2} & \longrightarrow & E_1^{0,2} \\ & & E_1^{-1,1} & \xrightarrow{d_2} & E_1^{0,1} \\ & & & & E_1^{0,0} \end{array}$$

coincide with

$$\begin{array}{ccccc} \Lambda^2 T^* & \xrightarrow{d_1} & \mathrm{CH}^1(X) \otimes T^* & \longrightarrow & \mathrm{CH}^2(X) \\ & & F^\times \otimes T^* & \xrightarrow{d_2} & F^\times \otimes \mathrm{CH}^1(X) \\ & & & & K_2(F) \end{array}$$

We would like to compute the differentials d_1 and d_2. Comparing with the case $n = 1$ we see immediately that $d_2 = \mathrm{id} \otimes d$, where d is as defined in the case $n = 1$.

In order to compute d_1 choose two characters $f, f' \in T^*$. By Proposition 5.3, we can find functions $h, h' \in F(X)[Y_\xi]^\times \subset F(G)^\times$ such that

$$h(tg) = f(t)h(g), \quad h'(tg) = f'(t)h'(g).$$

We may assume that the divisors $\operatorname{div}(h)$ and $\operatorname{div}(h')$ on G have no common components (if necessary modifying h by a function in $F(X)^\times$). Let

$$\operatorname{div}(h) = \pi^{-1}\Big(\sum_x n_x[x]\Big), \quad \operatorname{div}(h') = \pi^{-1}\Big(\sum_{x'} n_{x'}[x']\Big),$$

where x and x' range over the points of codimension 1 in X. The image of the symbol $\{h, h'\}$ under the residue homomorphism

$$A^0(\pi^{-1}\xi, K_2) \to \sum_{x \in X^{(1)}} A^0(\pi^{-1}x, K_1)$$

equals

$$\sum_x n_x[x, h'|_{\pi^{-1}x}] - \sum_{x'} n_{x'}[x', h|_{\pi^{-1}x'}].$$

Here by $h'|_{\pi^{-1}x}$ we mean the restriction of the function h' on the fiber $\pi^{-1}x$. We have $h'|_{\pi^{-1}x} \in F(x)[\pi^{-1}x]^\times$ if $n_x \neq 0$, as the divisors of h and h' are disjoint. The symbol $[x, h'|_{\pi^{-1}x}]$ denotes the corresponding element in

$$A^0(\pi^{-1}x, K_1) = F(x)[\pi^{-1}x]^\times.$$

The fiber $\pi^{-1}x$ is a principal homogeneous space of T over $F(x)$. The image of $h'|_{\pi^{-1}x}$ under

$$\kappa : F(x)[\pi^{-1}x]^{\times} \to T^*$$

is equal to f' and similarly, the image of $h|_{\pi^{-1}x'}$ equals f. Hence

$$d_1(f \wedge f') = \sum_x n_x[x] \otimes f' - \sum_{x'} n_{x'}[x'] \otimes f = d(f) \otimes f' - d(f') \otimes f.$$

Suppose now that G is a split simply connected semisimple group. Then d and d_2 are isomorphisms. If we identify $\mathrm{CH}^1(X)$ with T^* via d we see that d_1 is identified with the injective homomorphism

$$\Lambda^2 T^* \to T^* \otimes T^*, \quad f \wedge f' \mapsto f \otimes f' - f' \otimes f.$$

Hence, the cokernel of d_1 is isomorphic to the symmetric square $S^2(T^*)$. The spectral sequence then yields an isomorphism

$$K_2(F) \xrightarrow{\sim} A^0(G, K_2)$$

and an exact sequence

$$0 \longrightarrow A^1(G, K_2) \xrightarrow{\nu_G} S^2(T^*) \xrightarrow{\varsigma} \mathrm{CH}^2(X) \longrightarrow A^2(G, K_2).$$

By Proposition 6.1, $A^2(G, K_2) = \mathrm{CH}^2(G) = 0$. We have proved:

THEOREM 6.7. [EKLV 98, Prop. 3.20] *Let G be a split simply connected semisimple group, let $T \subset G$ be a split maximal torus, and set $X = G/T$. Then*

(1) *The natural map $K_2(F) \to A^0(G, K_2)$ is an isomorphism;*
(2) *The sequence*

$$0 \longrightarrow A^1(G, K_2) \xrightarrow{\nu_G} S^2(T^*) \xrightarrow{\varsigma} \mathrm{CH}^2(X) \longrightarrow 0$$

is exact.

We would like to study functorial properties of the map ν_G. Let $\varphi : G' \to G$ be a homomorphism of split simply connected semisimple groups and let $T \subset G$, $T' \subset G'$ be split maximal tori such that $\varphi(T') \subset T$. Set $X = G/T$ and $X' = G'/T'$. We have the projection $\pi' : G' \to X'$ and an induced morphism $\psi : X' \to X$.

We can view the cycle module M/X from 6.2 as a cycle module over X'. We have filtration (6.3) by cycle submodules over X' and isomorphisms (6.4) of cycle modules over X'. As above we get isomorphisms

$$A^{p+q}(X', M_n^{(p/p+1)}) \simeq K_{n-q}(F) \otimes \mathrm{CH}^{p+q}(X') \otimes \Lambda^{-p}\, T^*.$$

Similarly to M, we define the cycle module M' over X' via

$$M'_n(x') = A^0\left((\pi')^{-1}x', K_n\right)$$

for every point $x' \in X'(E)$. Considering M as a cycle module over X' we have a morphism of cycle modules $\alpha : M \to M'$ [Rost 96, p. 384]. By [Rost 96, Th. 12.7], the following diagram commutes, where φ^* and ψ^* are the pullback homomorphisms

[Rost 96, §12] and $\alpha_\sharp$ is induced by α.

$$
\begin{array}{ccccc}
A^1(G,K_2) & & \xrightarrow{\qquad\qquad\varphi^*\qquad\qquad} & & A^1(G',K_2) \\
\gamma_G \downarrow \wr & & & & \gamma_{G'} \downarrow \wr \\
A^1(X,M_2) & \xrightarrow{\psi^*} & A^1(X',M_2) & \xrightarrow{\alpha_\sharp} & A^1(X',M'_2) \\
\uparrow & & \uparrow & & \uparrow \\
A^1(X,M_2^{(-1)}) & \xrightarrow{\psi^*} & A^1(X',M_2^{(-1)}) & \xrightarrow{\alpha_\sharp} & A^1(X',M_2'^{(-1)}) \\
\downarrow & & \downarrow & & \downarrow \\
A^1(X,M_2^{(-1/0)}) & \xrightarrow{\psi^*} & A^1(X',M_2^{(-1/0)}) & \xrightarrow{\alpha_\sharp} & A^1(X',M_2'^{(-1/0)}) \\
\downarrow \wr & & \downarrow \wr & & \downarrow \wr \\
\mathrm{CH}^1(X)\otimes T^* & \xrightarrow{\psi^*\otimes\mathrm{id}} & \mathrm{CH}^1(X')\otimes T^* & \xrightarrow{\mathrm{id}\otimes\varphi^*} & \mathrm{CH}^1(X')\otimes T'^* \\
d\otimes\mathrm{id} \downarrow \wr & & d\otimes\mathrm{id} \downarrow \wr & & d\otimes\mathrm{id} \downarrow \wr \\
T^*\otimes T^* & \xrightarrow{\varphi^*\otimes\mathrm{id}} & T'^*\otimes T^* & \xrightarrow{\mathrm{id}\otimes\varphi^*} & T'^*\otimes T'^* \\
\downarrow & & & & \downarrow \\
S^2(T^*) & & \xrightarrow{\qquad\qquad S^2(\varphi^*)\qquad\qquad} & & S^2(T'^*)
\end{array}
$$

The homomorphism ν_G is defined via the sequence of homomorphisms

$$
A^1(G,K_2) \xrightarrow{\gamma_G} A^1(X,M_2) \twoheadleftarrow A^1(X,M_2^{(-1)}) \to A^1(X,M_2^{(-1/0)}) \xrightarrow{\sim} \\
\xrightarrow{\sim} \mathrm{CH}^1(X)\otimes T^* \xrightarrow{d\otimes\mathrm{id}} T^*\otimes T^* \to S^2(T^*),
$$

hence, commutativity of the diagram above yields:

PROPOSITION 6.8. *Let $\varphi : G' \to G$ be a homomorphism of split simply connected semisimple groups, and let $T \subset G$, $T' \subset G'$ be split maximal tori such that $\varphi(T') \subset T$. Then the diagram*

$$
\begin{array}{ccc}
A^1(G,K_2) & \xrightarrow{\nu_G} & S^2(T^*) \\
\varphi^* \downarrow & & \downarrow S^2(\varphi^*) \\
A^1(G',K_2) & \xrightarrow{\nu_{G'}} & S^2(T'^*)
\end{array}
$$

commutes.

LEMMA 6.9. (Cf. [EKLV 98, Lemma 4.3]) *Let G be a split semisimple simply connected group. Then the action of $G(F)$ on the K-cohomology groups $A^p(G,K_n)$ by left and right translations is trivial.*

PROOF. Let $U_\alpha \subset G$ be a root subgroup, and fix $u \in U_\alpha(F)$. Choose a morphism $h : \mathbf{A}^1 \to G$ such that $h(0) = 1$ and $h(1) = u$. Consider the morphisms

$$
s : \mathbf{A}^1 \times G \to G \quad \text{such that} \quad s(t,g) = h(t)\cdot g
$$

and

$$i_k : G \to \mathbf{A}^1 \times G \quad \text{where} \quad i_k(g) = (k, g), \quad k = 0, 1.$$

By the homotopy invariance property, the pullback map

$$p^* : A^p(G, K_n) \to A^p(\mathbf{A}^1 \times G, K_n)$$

with respect to the projection $p : \mathbf{A}^1 \times G \to G$ is an isomorphism. Since

$$p \circ i_0 = p \circ i_1 = \mathrm{id}_G = s \circ i_0,$$

we have $i_0^* = i_1^*$ and hence

$$\mathrm{id} = (s \circ i_0)^* = i_0^* \circ s^* = i_1^* \circ s^* = (s \circ i_1)^*.$$

The composition $s \circ i_1$ is the left translation by u, therefore, the translation action by u on $A^p(G, K_n)$ is trivial.

Since G is split simply connected, $G(F)$ is generated by the subgroups $U_\alpha(F)$ where α is a root, see for example [PR 85, 1.3, 1.6]. The result follows. □

6.10. A BASIS FOR $S^2(T^*)^W$. Let W be the Weyl group of a split simply connected group G. The group W acts naturally on T^* and $S^2(T^*)$. Let k be the number of connected components of the Dynkin diagram of G. Write $\Phi \subset T^*$ for the root system of G and $\Phi^\vee \subset T_*$ for the dual root system. The group $S^2(T^*)^W$ coincides with the group of all integral W-invariant quadratic forms on $T_* \otimes \mathbb{R}$, hence $S^2(T^*)^W$ is a free abelian group with canonical generators $q_1, q_2, \ldots, q_k$, where q_i is the (canonical) positive-definite W-invariant quadratic form which has value 1 on short coroots of the i-th component of $\Phi^\vee$ and is zero on other components.

6.11. THE GROUP $CH^2(X)$. As noticed in the proof of Lemma 6.5, the projection $X \to G/B$ induces an isomorphism on CH^2 and G/B is cellular. Hence $CH^2(X) \simeq CH^2(G/B)$ is a free group with basis the closures of the cells. (This is essentially proved in [Kahn 99, Lemma 3.3].) The cells in G/B are numbered by the elements of the Weyl group W as in the Bruhat decomposition, and the codimension of a cell is equal to the length of the corresponding element of W; that is, the rank of $\mathrm{CH}^2(X)$ is the number of length 2 elements of W. We have, where "vertices" and "edges" refer to the Dynkin diagram of G and $r = \dim(T)$:

$$\begin{aligned}
\operatorname{rank} \mathrm{CH}^2(X) &= \binom{\text{\# of pairs of non-}}{\text{connected vertices}} + 2\binom{\text{\# of pairs of connected}}{\text{vertices}} \\
&= (\text{\# of pairs of vertices}) + (\text{\# of edges}) \\
&= \binom{r}{2} + (r - k) = \frac{r(r+1)}{2} - k.
\end{aligned}$$

COROLLARY 6.12. *The homomorphism ν_G in Proposition 6.8 has image contained in $S^2(T^*)^W$. It induces an isomorphism $A^1(G, K_2) \xrightarrow{\sim} S^2(T^*)^W$ which is natural in G.*

PROOF. The first sentence follows from Proposition 6.8 where φ is conjugation by an element of the normalizer of T in G and Lemma 6.9.

Theorem 6.7.2 implies that

$$\ker\bigl(\zeta \colon S^2(T^*) \to \mathrm{CH}^2(X)\bigr) \subset S^2(T^*)^W. \tag{6.13}$$

We must show that this inclusion of free groups is actually an equality. Since $\mathrm{CH}^2(X)$ is free (6.11), the cokernel of the inclusion (6.13) is torsion-free. Therefore, it is sufficient to show that both groups in (6.13) have the same rank. We find:

$$\begin{aligned}\operatorname{rank}\ker(\zeta) &= \operatorname{rank} S^2(T^*) - \operatorname{rank}\mathrm{CH}^2(X)\\ &= \frac{r(r+1)}{2} - \left(\frac{r(r+1)}{2} - k\right) = k\\ &= \operatorname{rank} S^2(T^*)^W.\end{aligned}$$

□

EXAMPLE 6.14. Consider the simplest case $G = \mathbf{SL}_2$. Let T be the torus of diagonal matrices. The character $f \in T^*$, $f(t) = t_{11}$ is a generator of T^*. The functions x_{11} and x_{12} on G satisfy $x_{1i}(tg) = f(t)x_{1i}(g)$, $i = 1, 2$. Hence, the image of the cycle class $[x_{11} = 0, x_{12}] \in A^1(G, K_2)$ in $S^2(T^*)$ is the canonical generator $f^2 = f \cdot f$ in $S^2(T^*)$.

If G is simple, the group $S^2(T^*)^W$ is infinite cyclic with the canonical generator q_G taking value 1 on all short coroots of G.

EXAMPLE 6.15. Let $G = \mathbf{SL}_n$, $n \geq 2$. The split maximal torus T of diagonal matrices of G is isomorphic to the kernel of the product homomorphism

$$(\mathbf{G}_{\mathrm{m}})^n \to \mathbf{G}_{\mathrm{m}}\,.$$

Hence the group of cocharacters T_* can be identified with the subgroup in $\mathbb{Z}^n$ consisting of all n-tuples $x = (x_1, x_2, \dots, x_n)$ such that the sum of the x_i is 0 [KMRT 98, §24]. The Weyl group $W = S_n$ acts by permuting the x_i. Clearly, the W-invariant integral quadratic form

$$q_G(x) = \frac{1}{2}\sum_{i=1}^{n} x_i^2 = -\sum_{i<j} x_i x_j$$

is the canonical generator of $S^2(T^*)^W$. It takes value 1 on the (short) coroots $\pm(e_i - e_j)$ for $i \neq j$.

7. The group $Q(G)$

In this section we define the group $Q(G)$ for every simply connected group G (see [KMRT 98, §31]) and prove that it is canonically isomorphic to $A^1(G, K_2)$ if G is split. For a generalization see Theorem 9.3.

Let G be an algebraic group over a separably closed field F. A *loop* in G is a group homomorphism $\mathbf{G}_{\mathrm{m}} \to G$ over F. We write G_* for the set of all loops in G. In general there is no group structure on G_*, but if $f_1, f_2, \dots, f_r$ are loops with pairwise commuting images, then the pointwise product $f_1 f_2 \cdots f_r$ is also a loop. In particular, for an integer n and any loop f the n-th power f^n is defined. For every $g \in G(F)$ and any loop f, we write ${}^g f$ for the conjugate loop

$$\mathbf{G}_{\mathrm{m}} \xrightarrow{\ f\ } G \xrightarrow{\mathrm{Inn}(g)} G,$$

where $\mathrm{Inn}(g)$ is given by $\mathrm{Inn}(g)(x) = gxg^{-1}$. Consider the abelian group $Q(G)$ of all functions $q : G_* \to \mathbb{Z}$ such that

(1) $q({}^g f) = q(f)$ for all $g \in G(F)$ and $f \in G_*$,

(2) for all loops $f_1, f_2, \dots, f_r$ with pairwise commuting images, the function

$$\mathbb{Z}^r \to \mathbb{Z}, \quad (k_1, k_2, \dots, k_r) \mapsto q(f_1^{k_1} f_2^{k_2} \dots f_r^{k_r})$$

is a quadratic form on $\mathbb{Z}^r$.

For an algebraic group G over an arbitrary field F the absolute Galois group $\mathrm{Gal}(F)$ acts naturally on $Q(G_{\mathrm{sep}})$ and we set

$$Q(G) = Q(G_{\mathrm{sep}})^{\mathrm{Gal}(F)}.$$

EXAMPLE 7.1. Let T be a split torus. Then T_* coincides with the group of cocharacters of T and $Q(T)$ is canonically isomorphic to the group $S^2(T^*)$ of integral quadratic forms on the lattice T_*.

The operation Q has nice functorial properties.

(1) Every group homomorphism $\alpha : G \to G'$ induces a group homomorphism

$$Q(\alpha)\colon Q(G') \to Q(G), \quad Q(\alpha)(q)(\beta) = q(\alpha \circ \beta).$$

(2) For a field extension E/F there is the *restriction homomorphism*

$$\mathrm{Res}_{E/F} : Q(G) \to Q(G_E).$$

If F is algebraically closed in E, $\mathrm{Res}_{E/F}$ is an isomorphism.

(3) For a finite separable field extension E/F there is the *corestriction homomorphism*

$$\mathrm{Cor}_{E/F} : Q(G_E) \to Q(G).$$

Let G be a reductive group over F and let $T \subset G$ be a maximal torus over F. Let W denote the Weyl group of G_{sep} relative to T_{sep}. The image of the $\mathrm{Gal}(F)$-equivariant map

$$Q(G_{\mathrm{sep}}) \to Q(T_{\mathrm{sep}}) = S^2(T_{\mathrm{sep}}^*)$$

consists of W-invariant elements. Thus, we get a homomorphism

$$\beta : Q(G) \to \left(S^2(T_{\mathrm{sep}}^*)^W\right)^{\mathrm{Gal}(F)}.$$

PROPOSITION 7.2. *The map β is an isomorphism.*

PROOF. We can assume that F is separably closed. The image of a loop $f : \mathbf{G}_{\mathrm{m}} \to G$ is either trivial or is isomorphic to $\mathbf{G}_{\mathrm{m}}$ and hence is conjugate by an element in $G(F)$ to a subgroup of T. This implies that β is injective. To show surjectivity of β, choose a quadratic form $q \in S^2(T_{\mathrm{sep}}^*)^W$. We would like to extend q to a quadratic map $\tilde{q}$ on all the loops in G. Let $f : \mathbf{G}_{\mathrm{m}} \to G$ be a loop. We choose a split maximal torus $S \subset G$ containing the image of f. By [Bor 91, 20.9(ii)], there is an element $g \in G(F)$ such that ${}^gS = T$, where ${}^gS = \mathrm{Inn}(g)(S)$. Set $\tilde{q}(f) = q({}^gf)$.

We need to prove that $\tilde{q}$ is well-defined. Let S' be another split maximal torus containing $\mathrm{im}(f)$ and let $g' \in G(F)$ be such that ${}^{g'}S' = T$. The centralizer H of $\mathrm{im}(f)$ is a reductive group with split maximal tori S and S' [Hu 75, Th. 34.4(c)]. Again by [Bor 91, 20.9(ii)], the tori S and S' of H are conjugate by an element of $H(F)$, i.e., $S' = {}^hS$ for some $h \in H(F)$. Then $g'h = wg$, where w belongs to the normalizer of T in G. Finally,

$$q({}^{g'}f) = q({}^{g'h}f) = q({}^{wg}f) = q({}^gf). \qquad \square$$

Let G be a split simply connected simple algebraic group and fix $T \subset G$ a split maximal torus. The Galois group $\mathrm{Gal}(F)$ acts trivially on T^*_{sep}, hence β is an isomorphism between $Q(G)$ and $S^2(T^*)^W$. We will denote the composition

$$A^1(G, K_2) \xrightarrow{\nu_G} S^2(T^*)^W \xrightarrow{\beta^{-1}} Q(G)$$

by μ_G. Since all split maximal tori in G are conjugate by an element of $G(F)$, Proposition 6.8 and Lemma 6.9 show that μ_G does not depend on the choice of T. Proposition 6.12 then yields:

PROPOSITION 7.3. *Let G be a split simply connected simple group. Then $\mu_G\colon A^1(G, K_2) \xrightarrow{\sim} Q(G)$ is an isomorphism which is natural in G.*

We can give a direct description of μ_G. Let

$$p : A^1(\mathbf{SL}_2, K_2) \xrightarrow{\sim} \mathbb{Z}$$

be the isomorphism taking the generator $[x_{11} = 0, x_{12}]$ to 1 (see Example 6.14). Every coroot $\alpha : \mathbf{G}_{\mathrm{m}} \to G$ can be extended to a group homomorphism $\bar{\alpha} : \mathbf{SL}_2 \to G$.

PROPOSITION 7.4. *For every coroot α of G and $v \in A^1(G, K_2)$, we have $\mu_G(v)(\alpha) = p\big(\bar{\alpha}^*(v)\big)$.*

PROOF. The proposition follows from the commutativity of the diagram (Proposition 7.3):

$$\begin{array}{ccccc} A^1(G, K_2) & \xrightarrow{\mu_G} & Q(G) & & \\ \Big\downarrow{\scriptstyle \bar{\alpha}^*} & & \Big\downarrow{\scriptstyle Q(\bar{\alpha})} & & \\ A^1(\mathbf{SL}_2, K_2) & \xrightarrow{\mu_{\mathbf{SL}_2}} & Q(\mathbf{SL}_2) & = & \mathbb{Z}. \\ \multicolumn{5}{c}{\xrightarrow[p]{\hspace{14em}}} \end{array}$$

□

REMARK 7.5. A quadratic function in $Q(G)$ is uniquely determined by its values on all coroots of G. Hence, the formula in Proposition 7.4 determines the map μ_G.

Let $T \subset G$ be a maximal torus defined over F of a simply connected group G. As noticed in 6.10, $S^2(T^*_{\mathrm{sep}})^W$ is a free abelian group with canonical generators $q_1, q_2, \dots, q_k$, where k is the number of connected components of the root system Φ, the form q_i has value 1 on short coroots of the i-th component of $\Phi^\vee$ and is zero on other components. The Galois group $\mathrm{Gal}(F)$ permutes the q_i in accordance with the action on the set of irreducible components of Φ. In particular, if G is absolutely simple, the group $Q(G)$ is infinite cyclic with canonical generator q_G taking value 1 on all short coroots (loops) of G_{sep}.

The following properties of the Q-operation on simply connected groups can be derived from Proposition 7.2:

PROPOSITION 7.6. *For simply connected groups G and G' the natural homomorphisms from G and G' to $G \times G'$ induce an isomorphism*

$$Q(G \times G') \xrightarrow{\sim} Q(G) \oplus Q(G').$$

Let E/F be a finite separable field extension, and let G be an algebraic group over E. The *corestriction of G from E to F* [KMRT 98, 20.5] is a group scheme $R_{E/F}(G)$ over F such that for every commutative F-algebra R,

$$R_{E/F}(G)(R) = G(R \otimes_F E).$$

PROPOSITION 7.7. *Let E/F be a finite separable field extension, let G be a simply connected group over E, and set $G' = R_{E/F}(G)$. There are canonical group homomorphisms $i : G'_E \to G$ and $j : G \to G'_E$. The compositions*

$$Q(G') \xrightarrow{\text{Res}_{E/F}} Q(G'_E) \xrightarrow{Q(j)} Q(G)$$

and

$$Q(G) \xrightarrow{Q(i)} Q(G'_E) \xrightarrow{\text{Cor}_{E/F}} Q(G')$$

are isomorphisms which are inverse to each other.

Propositions 7.6 and 7.7 yield:

PROPOSITION 7.8. *For $j = 1, \ldots, s$, let E_j/F be a finite separable extension and let G_j be an absolutely simple and simply connected group over E_j. There is a canonical isomorphism*

$$Q\Big(\prod_{j=1}^{s} R_{E_j/F}(G_j)\Big) \simeq \coprod_{j=1}^{s} Q(G_j).$$

By [Kne 69, pp. 25, 26] or [Ti 66], every semisimple simply connected group G over F can be written as such a product $\prod R_{E_j/F}(G_j)$, hence the group $Q(G)$ is a free abelian group of rank s with canonical basis whose elements correspond to the canonical generators q_{G_j} under this isomorphism.

Let $\alpha : G \to G'$ be a homomorphism of absolutely simple simply connected groups. Then

$$Q(\alpha) : Q(G') \to Q(G)$$

is a homomorphism of infinite cyclic groups with generators $q_{G'}$ and q_G. Hence

$$Q(\alpha)(q_{G'}) = n_\alpha \cdot q_G$$

for a uniquely determined integer n_α. We call n_α the *Rost multiplier* of α; it is sometimes called the "Dynkin index" of α.

In the following proposition we collect some properties of the Rost multipliers.

PROPOSITION 7.9. *Let $\alpha : G \to G'$ be a homomorphism of absolutely simple simply connected groups.*

(1) *For a homomorphism $\beta : G' \to G''$ of absolutely simple simply connected groups, $n_{\beta\circ\alpha} = n_\beta \cdot n_\alpha$.*
(2) *If $\beta : \mathbf{G}_{\mathrm{m}} \to G_{\mathrm{sep}}$ is a short coroot of G, then $n_\alpha = q_{G'}(\alpha \circ \beta)$.*
(3) *The integer n_α is non-negative and $n_\alpha = 0$ if and only if α is trivial.*
(4) *The integer n_α does not change under field extensions.*
(5) *Let $\rho : G \to \mathbf{SL}(V)$ and $\mu : G \to \mathbf{SL}(W)$ be two representations. Then*
    (a) *$n_{\rho\oplus\mu} = n_\rho + n_\mu$ and*
    (b) *$n_{\rho\otimes\mu} = \dim(W) \cdot n_\rho + \dim(V) \cdot n_\mu$.*

Equation 7.9.5b is known as the *derivation formula.*

PROOF. (1): Obvious.
(2): Since $q_G(\beta) = 1$, we have $n_\alpha = n_\alpha q_G(\beta) = Q(\alpha)(q_{G'})(\beta) = q_{G'}(\alpha \circ \beta)$.
(3): Follows from (2) and the fact that the form $q_{G'}$ is positive-definite.
(4): Follows from (2) and the fact that the canonical generator $q_{G'}$ is preserved under field extensions.

(5): The representation $\rho \oplus \mu$ is the composition

$$\rho \oplus \mu : G \xrightarrow{(\rho,\mu)} \mathbf{SL}(V) \times \mathbf{SL}(W) \xrightarrow{\nu} \mathbf{SL}(V \oplus W),$$

where ν is the natural inclusion. The map on the Q-groups with respect to the natural inclusion of $\mathbf{SL}(V)$ (resp. $\mathbf{SL}(W)$) into $\mathbf{SL}(V \oplus W)$ takes $q_{\mathbf{SL}(V\oplus W)}$ to $q_{\mathbf{SL}(V)}$ (resp. $q_{\mathbf{SL}(W)}$, see Example 7.10). Hence,

$$Q(\nu)(q_{\mathbf{SL}(V\oplus W)}) = q_{\mathbf{SL}(V)} + q_{\mathbf{SL}(W)}$$

and

$$\begin{aligned} Q(\rho \oplus \mu)(q_{\mathbf{SL}(V\oplus W)}) &= Q(\rho,\mu) \circ Q(\nu)(q_{\mathbf{SL}(V\oplus W)}) \\ &= Q(\rho,\mu)(q_{\mathbf{SL}(V)} + q_{\mathbf{SL}(W)}) \\ &= n_\rho \cdot q_G + n_\mu \cdot q_G. \end{aligned}$$

The representation $\rho \otimes \mu$ is the composition

$$\rho \oplus \mu : G \xrightarrow{(\rho,\mu)} \mathbf{SL}(V) \times \mathbf{SL}(W) \xrightarrow{\varepsilon} \mathbf{SL}(V \otimes W),$$

where ε is the natural inclusion. The restriction of ε on $\mathbf{SL}(V)$ (resp. $\mathbf{SL}(W)$) is isomorphic to the representation $\dim W \cdot \mathrm{id}_{\mathbf{SL}(V)}$ (resp. $\dim V \cdot \mathrm{id}_{\mathbf{SL}(W)}$). Hence,

$$Q(\varepsilon)(q_{\mathbf{SL}(V\otimes W)}) = \dim W \cdot q_{\mathbf{SL}(V)} + \dim V \cdot q_{\mathbf{SL}(W)}$$

and

$$\begin{aligned} Q(\rho \otimes \mu)(q_{\mathbf{SL}(V\otimes W)}) &= Q(\rho,\mu) \circ Q(\varepsilon)(q_{\mathbf{SL}(V\otimes W)}) \\ &= Q(\rho,\mu)(\dim W \cdot q_{\mathbf{SL}(V)} + \dim V \cdot q_{\mathbf{SL}(W)}) \\ &= \dim W \cdot n_\rho \cdot q_G + \dim V \cdot n_\mu \cdot q_G. \end{aligned}$$ □

EXAMPLE 7.10. For the standard inclusion $\rho : \mathbf{SL}_n \hookrightarrow \mathbf{SL}_m$ $(m > n)$ we have $n_\rho = 1$ since the coroots of $\mathbf{SL}_n$ are also coroots of $\mathbf{SL}_m$ and have the same length.

EXAMPLE 7.11. Let $\rho : \mathbf{Sp}_{2n} \hookrightarrow \mathbf{SL}_{2n}$ be the standard embedding. The embedding of maximal tori is defined by

$$(t_1, t_2, \dots, t_n) \mapsto (t_1, t_2, \dots, t_n, t_1^{-1}, t_2^{-1}, \dots, t_n^{-1}).$$

Hence, the map of cocharacter groups takes $(x_1, x_2, \dots, x_n)$ to

$$(x_1, x_2, \dots, x_n, -x_1, -x_2, \dots, -x_n).$$

The image of the short coroot $\pm e_i$ is the short coroot $\pm(e_i - e_{n+i})$, hence $n_\rho = 1$.

EXAMPLE 7.12. Let $\rho : \mathbf{Spin}_{2n} \hookrightarrow \mathbf{Spin}_{2n+1}$ be the standard embedding with $n \geq 3$. A maximal torus of the first group is also maximal in the second. The short coroots $\pm e_i \pm e_j$ of G' correspond to the same short coroots of G, hence $n_\rho = 1$.

EXAMPLE 7.13. Let $\rho : \mathbf{Spin}_{2n+1} \hookrightarrow \mathbf{Spin}_{2n+2}$ be the standard embedding with $n \geq 2$. The homomorphism of cocharacter groups of maximal tori is induced by the canonical inclusion $\mathbb{Z}^n \hookrightarrow \mathbb{Z}^{n+1}$. The cocharacters $\pm e_i \pm e_j$ are the short coroots in both the groups, hence $n_\rho = 1$.

EXAMPLE 7.14. Let $\rho : \mathbf{Spin}_n \hookrightarrow \mathbf{Spin}_m$ be the standard embedding with $m > n \geq 5$. By Examples 7.12 and 7.13, $n_\rho = 1$.

EXAMPLE 7.15. Let ρ be the composition

$$\mathbf{Spin}_{2n} \xrightarrow{\alpha} \mathbf{O}_{2n}^{+} \hookrightarrow \mathbf{SL}_{2n},$$

where α is the standard isogeny and $n \geq 3$. The cocharacter group of the maximal torus of $\mathbf{Spin}_{2n}$ is contained in $\mathbb{Z}^n$ with the coroots $\pm e_i \pm e_j$ of the same length. The corresponding cocharacter of the maximal torus of $\mathbf{SL}_{2n}$ is $\pm(f_i - g_i) \pm (f_j - g_j)$ if we identify the group of all cocharacters with a subgroup in $\coprod \mathbb{Z} f_i \oplus \coprod \mathbb{Z} g_i$. By Example 6.15 and Proposition 7.9.2, $n_\rho = 2$.

8. A theorem of B. Kahn

In this section we prove a theorem of B. Kahn which relates the group $A^1(G, K_2)$ for a simply connected group G over F to the group $H^3\big(F, \mathbb{Q}/\mathbb{Z}(2)\big)$.

We write $\mathbf{S}(X)$ for the category of étale sheaves of abelian groups over a scheme X [Milne 80, Ch. 2]. The functor of global sections

$$f_X : \mathbf{S}(X) \to \textbf{Abelian Groups}, \quad P \mapsto P(X)$$

gives rise to the right derived functor

$$Rf_X : D^+\big(\mathbf{S}(X)\big) \to D^+\big(\textbf{Abelian Groups}\big),$$

where D^+ stands for the derived category of bounded below complexes [GM 99, Ch. 3]. If C is in $D^+\big(\mathbf{S}(X)\big)$, we write $\mathbb{H}^n_{ét}(X, C)$ for the cohomology group $H^n\big(Rf_X(C)\big)$.

In [Li 87] and [Li 90] S. Lichtenbaum defined a (weight two motivic) complex $\Gamma(2) = \Gamma(2)_X$ of étale sheaves in $\mathbf{S}(X)$ for every scheme X. The complex $\Gamma(2)$ is supported in degrees 1 and 2. If X is a smooth irreducible variety over F, we have ([Li 90], [Kahn 96, Th. 1.1]):

$$\mathbb{H}^q_{ét}\big(X, \Gamma(2)\big) \simeq \begin{cases} 0 & \text{if } q \leq 0; \\ K_3\big(F(X)\big)_{\text{ind}} & \text{if } q = 1; \\ A^0(X, K_2) & \text{if } q = 2; \\ A^1(X, K_2) & \text{if } q = 3, \end{cases} \tag{8.1}$$

where we write $K_3(E)_{\text{ind}}$ for the cokernel of the natural homomorphism $K_3(E) \to K_*^Q(E)$ (here E is a field and K_*^Q is the Quillen's K-theory functor [Qu 73]) and an exact sequence

$$0 \to \mathrm{CH}^2(X) \to \mathbb{H}^4_{ét}\big(X, \Gamma(2)\big) \to H^3\big(F(X), \mathbb{Q}/\mathbb{Z}(2)\big). \tag{8.2}$$

Write $\mathbb{H}^q_{ét}\big(F, \Gamma(2)\big)$ for $\mathbb{H}^q_{ét}\big(\operatorname{Spec} F, \Gamma(2)\big)$. We have in particular:

$$\mathbb{H}^q_{ét}\big(F, \Gamma(2)\big) \simeq \begin{cases} 0 & \text{if } q \leq 0; \\ K_3(F)_{\text{ind}} & \text{if } q = 1; \\ K_2(F) & \text{if } q = 2; \\ 0 & \text{if } q = 3. \end{cases} \tag{8.3}$$

For every n prime to $\operatorname{char} F$, there is a distinguished triangle in the category $D^+\big(\mathbf{S}(\operatorname{Spec} F)\big)$ [Li 87, Prop. 3.3]:

$$\Gamma(2)_F \xrightarrow{n} \Gamma(2)_F \to \mu_n^{\otimes 2} \to \Gamma(2)_F[1]$$

and when $p = \operatorname{char} F > 0$, a triangle [Li 90, Th. 1.1(c)]

$$\Gamma(2)_F \xrightarrow{p^m} \Gamma(2)_F \to \nu_m(2)[-2] \to \Gamma(2)_F[1],$$

where $\nu_m(2) = K_2(F_{\text{sep}})/p^m K_2(F_{\text{sep}})$. These triangles together with (A.4) and the fact that $\mathbb{H}^4_{ét}(F, \Gamma(2))$ is a torsion group yield an isomorphism [Kahn 93, Lemme 1.1]

$$\mathbb{H}^4_{ét}(F, \Gamma(2)) \simeq H^3(F, \mathbb{Q}/\mathbb{Z}(2)). \tag{8.4}$$

Let $p : X \to \operatorname{Spec} F$ be the structure morphism of a scheme X. Since the complex $\Gamma(2)_X$ is natural in X, there is a canonical morphism

$$\Gamma(2)_F \to Rp_*\Gamma(2)_X$$

in $D^+(\mathbf{S}(\operatorname{Spec} F))$. We write $C(X)$ for the cone of this morphism and set

$$\overline{\mathbb{H}}^n_{ét}(X, \Gamma(2)) = \mathbb{H}^n_{ét}(F, C(X)) = H^n(Rf_F(C(X))).$$

Applying the functor Rf to the distinguished triangle

$$\Gamma(2)_F \to Rp_*\Gamma(2)_X \to C(X) \to \Gamma(2)_F[1]$$

in $D^+(\mathbf{S}(\operatorname{Spec} F))$ we get the triangle

$$Rf_F(\Gamma(2)_F) \to Rf_X(\Gamma(2)_X) \to Rf_F(C(X)) \to Rf_F(\Gamma(2)_F)[1]$$

in D^+(**Abelian Groups**) that induces a long exact sequence

$$\to \mathbb{H}^n_{ét}(F, \Gamma(2)) \to \mathbb{H}^n_{ét}(X, \Gamma(2)) \to \overline{\mathbb{H}}^n_{ét}(X, \Gamma(2)) \to \mathbb{H}^{n+1}_{ét}(F, \Gamma(2)) \to \tag{8.5}$$

LEMMA 8.6. *Let X be smooth irreducible variety over F. Then*

$$\overline{\mathbb{H}}^q_{ét}(X, \Gamma(2)) = \begin{cases} 0 & \text{if } q \le 0; \\ \operatorname{coker}[K_3(F)_{\text{ind}} \to K_3(F(X))_{\text{ind}}] & \text{if } q = 1; \\ \operatorname{coker}[K_2(F) \to A^0(X, K_2)] & \text{if } q = 2. \end{cases}$$

PROOF. The statement follows from (8.1), (8.3), (8.5) and injectivity of the natural homomorphisms

$$K_3(F)_{\text{ind}} \to K_3(F(X))_{\text{ind}} \quad \text{and} \quad K_2(F) \to K_2(F(X))$$

[MS 90, Prop. 11.3], [Su 84, Th. 14.3]. □

PROPOSITION 8.7. *Let X be smooth irreducible variety over F such that the natural homomorphism $K_2(F_{\text{sep}}) \to A^0(X_{\text{sep}}, K_2)$ is an isomorphism. Then*

$$\overline{\mathbb{H}}^3_{ét}(X, \Gamma(2)) \simeq A^1(X_{\text{sep}}, K_2)^{\operatorname{Gal}(F)}.$$

PROOF. It follows from (8.1) and (8.5) applied to the variety X_{sep} over F_{sep} that

$$\overline{\mathbb{H}}^3_{ét}(X_{\text{sep}}, \Gamma(2)) \simeq A^1(X_{\text{sep}}, K_2). \tag{8.8}$$

Consider the Leray spectral sequence associated with the morphism $p : X \to \operatorname{Spec} F$ [Milne 80, Th. III.1.18]:

$$E_2^{p,q} = H^p(F, \overline{\mathbb{H}}^q_{ét}(X_{\text{sep}}, \Gamma(2))) \Longrightarrow \overline{\mathbb{H}}^{p+q}_{ét}(X, \Gamma(2)).$$

By [MS 90, Prop. 11.6], the cokernel of the natural homomorphism $K_3(F_{\text{sep}})_{\text{ind}} \to K_3(F_{\text{sep}}(X))_{\text{ind}}$ is uniquely divisible. Hence, it follows from Lemma 8.6 that $E_2^{p,q} = 0$ for $p \ge 1$ and $q \le 2$. Therefore, the spectral sequence together with (8.8) yield an isomorphism

$$\overline{\mathbb{H}}^3_{ét}(X, \Gamma(2)) \simeq E_\infty^{0,3} = E_2^{0,3} = A^1(X_{\text{sep}}, K_2)^{\operatorname{Gal}(F)}.$$ □

Collecting together (8.1)–(8.5) and Proposition 8.7 we get:

THEOREM 8.9. [Kahn 96, Th. 1] *Let X be a smooth irreducible variety over F. Suppose that the canonical map $K_2(F_{\text{sep}}) \to A^0(X_{\text{sep}}, K_2)$ is an isomorphism. Then the sequence*

$$0 \to A^1(X, K_2) \xrightarrow{\iota} A^1(X_{\text{sep}}, K_2)^{\text{Gal}(F)} \xrightarrow{d_X}$$
$$\to \ker\left[H^3(F, \mathbb{Q}/\mathbb{Z}(2)) \to H^3(F(X), \mathbb{Q}/\mathbb{Z}(2))\right] \to \text{CH}^2(X)$$

is exact, where the map ι is the natural one.

REMARK 8.10. By the construction, the homomorphism d_X in Theorem 8.9 is natural in X, i.e., for a morphism $g : Y \to X$ of varieties which satisfy the conditions of Theorem 8.9 the homomorphism d_X coincides with the composition of d_Y and the pullback homomorphism

$$g^* : A^1(X_{\text{sep}}, K_2)^{\text{Gal}(F)} \to A^1(Y_{\text{sep}}, K_2)^{\text{Gal}(F)}.$$

9. Definition of the Rost invariant

In this section we define the Rost invariant of an absolutely simple simply connected group.

Let G be a simply connected semisimple group over a field F; let Y be a principal homogeneous space of G over F. Choose an isomorphism $\alpha : G_{\text{sep}} \xrightarrow{\sim} Y_{\text{sep}}$ of principal homogeneous spaces.

LEMMA 9.1. *The pullback isomorphism $\alpha^* : A^p(Y_{\text{sep}}, K_n) \xrightarrow{\sim} A^p(G_{\text{sep}}, K_n)$ is $\text{Gal}(F)$-equivariant and does not depend on the choice of α.*

PROOF. Another choice β of an isomorphism differs from α by translation by an element of $G(F_{\text{sep}})$. By Lemma 6.9, $\beta^* = \alpha^*$. The same reasoning shows that $\gamma \circ \alpha^* = \alpha^* \circ \gamma$ for every $\gamma \in \text{Gal}(F)$. □

We write μ_Y for the composition

$$A^1(Y, K_2) \to A^1(Y_{\text{sep}}, K_2)^{\text{Gal}(F)} \xrightarrow{\alpha^*}$$
$$\to A^1(G_{\text{sep}}, K_2)^{\text{Gal}(F)} \xrightarrow{\mu_{G\text{sep}}} Q(G_{\text{sep}})^{\text{Gal}(F)} = Q(G).$$

By Theorem 6.7.1,

$$A^0(Y_{\text{sep}}, K_2) = K_2(F_{\text{sep}})$$

since $Y_{\text{sep}} \simeq G_{\text{sep}}$. Hence we can apply Theorem 8.9 to the variety Y. In particular, we get a homomorphism

$$d_Y : A^1(Y_{\text{sep}}, K_2)^{\text{Gal}(F)} \to H^3(F, \mathbb{Q}/\mathbb{Z}(2)).$$

We write

$$\delta_Y : Q(G) \to H^3(F, \mathbb{Q}/\mathbb{Z}(2))$$

for the composition of $d_Y \circ (\mu_{G\text{sep}} \circ \alpha^*)^{-1}$. Since $\text{CH}^2(Y) = 0$ by Proposition 6.1, Theorem 8.9 yields:

PROPOSITION 9.2. *The sequence*

$$0 \to A^1(Y, K_2) \xrightarrow{\mu_Y} Q(G) \xrightarrow{\delta_Y} H^3(F, \mathbb{Q}/\mathbb{Z}(2)) \to H^3(F(Y), \mathbb{Q}/\mathbb{Z}(2))$$

is exact.

We can now generalize Proposition 7.3. We apply Proposition 9.2 for $Y = G$.

THEOREM 9.3. *Let G be a (not necessarily split) simply connected group. Then the map*

$$\mu_G : A^1(G, K_2) \to Q(G)$$

is an isomorphism. For every group homomorphism $\varphi : G' \to G$ of simply connected groups, the following diagram commutes:

$$\begin{array}{ccc} A^1(G, K_2) & \xrightarrow[\mu_G]{\sim} & Q(G) \\ {\scriptstyle \varphi^*}\downarrow & & \downarrow{\scriptstyle Q(\varphi)} \\ A^1(G', K_2) & \xrightarrow[\mu_{G'}]{\sim} & Q(G'). \end{array}$$

PROOF. Note that the map on the H^3-groups in Proposition 9.2 is injective since G has a rational point (Proposition A.10). The second statement follows from Proposition 7.3 and the definition of μ_G. □

Fix a quadratic form $q \in Q(G)$ for a simply connected group G over F. The assignment

$$H^1(E, G) \to H^3\big(E, \mathbb{Q}/\mathbb{Z}(2)\big), \quad Y \mapsto \delta_Y(q_E)$$

for every field extension E/F gives rise to a normalized invariant of G with values in $H^3\big(\mathbb{Q}/\mathbb{Z}(2)\big)$, which we denote by $r_G(q)$. Thus, we have a group homomorphism

$$r_G : Q(G) \to \operatorname{Inv}^3\big(G, \mathbb{Q}/\mathbb{Z}(2)\big)_{\text{norm}}.$$

The following Proposition is a consequence of Theorem 9.3 and Remark 8.10:

PROPOSITION 9.4. *For every homomorphism $\varphi : G' \to G$ of simply connected groups the following diagram commutes:*

$$\begin{array}{ccc} Q(G) & \xrightarrow{r_G} & \operatorname{Inv}^3\big(G, \mathbb{Q}/\mathbb{Z}(2)\big)_{\text{norm}} \\ {\scriptstyle Q(\varphi)}\downarrow & & \downarrow{\scriptstyle \varphi^*} \\ Q(G') & \xrightarrow{r_{G'}} & \operatorname{Inv}^3\big(G', \mathbb{Q}/\mathbb{Z}(2)\big)_{\text{norm}}. \end{array}$$

Consider an injective homomorphism $\rho : G \hookrightarrow S$ into a special group S and the classifying variety $X = S/\rho(G)$. The canonical morphism

$$\pi : S \to X$$

is a G-torsor. Let Y_ξ be the generic fiber of the morphism π. It is a principal homogeneous space of G over the field $F(X)$. The function field $F(X)(Y_\xi)$ is naturally isomorphic to $F(S)$. Since F is algebraically closed in $F(X)$ the natural map $j : Q(G) \to Q\big(G_{F(X)}\big)$ is an isomorphism. We write τ for the composition

$$A^1(Y_\xi, K_2) \xrightarrow{\mu_{Y_\xi}} Q\big(G_{F(X)}\big) \xrightarrow{j^{-1}} Q(G).$$

PROPOSITION 9.5. *The sequence*

$$0 \longrightarrow A^1(Y_\xi, K_2) \xrightarrow{\tau} Q(G) \xrightarrow{r_G} \operatorname{Inv}^3\big(G, \mathbb{Q}/\mathbb{Z}(2)\big)_{\text{norm}} \longrightarrow 0$$

is exact.

PROOF. In the commutative diagram

$$\begin{array}{ccccccc} & & Q(G) & \xrightarrow{r_G} & \operatorname{Inv}^3(G, \mathbb{Q}/\mathbb{Z}(2))_{\mathrm{norm}} & & \\ & & {\scriptstyle j}\downarrow\wr & & \downarrow{\scriptstyle \theta_G} & & \\ 0 \longrightarrow A^1(Y_\xi, K_2) & \xrightarrow{\mu_{Y_\xi}} & Q(G_{F(X)}) & \xrightarrow{\delta_{Y_\xi}} & H^3(F(X), \mathbb{Q}/\mathbb{Z}(2)) & \xrightarrow{i} & H^3(F(S), \mathbb{Q}/\mathbb{Z}(2)) \end{array}$$

the bottom row is exact by Proposition 9.2 and θ_G is injective by Theorem 3.3. The principal homogeneous space Y_ξ is trivial over $F(S)$, hence we have $i \circ \theta_G = 0$. The statement follows by diagram chase. □

Write $Q(G)_0$ for the kernel of r_G.

COROLLARY 9.6. *The homomorphism τ induces an isomorphism*

$$\tau_0 : A^1(Y_\xi, K_2) \xrightarrow{\sim} Q(G)_0.$$

We study now some functorial properties of the invariants.

PROPOSITION 9.7. *Let G and G' be simply connected groups. Then the natural homomorphism*

$$\operatorname{Inv}^3\big(G, \mathbb{Q}/\mathbb{Z}(2)\big)_{\mathrm{norm}} \oplus \operatorname{Inv}^3\big(G', \mathbb{Q}/\mathbb{Z}(2)\big)_{\mathrm{norm}} \to \operatorname{Inv}^3\big(G \times G', \mathbb{Q}/\mathbb{Z}(2)\big)_{\mathrm{norm}}$$

is an isomorphism.

PROOF. The map is injective since it has a splitting induced by the embeddings of G and G' into $G \times G'$. Surjectivity follows from the surjectivity of r_G and Proposition 7.6. □

Let L/F be a separable field extension, and let G be an algebraic group over L. Then the corestriction $G' = R_{L/F}(G)$ is an algebraic group over F. By the Shapiro-Faddeev lemma, the assignment $Y \mapsto R_{L/F}(Y)$ provides a bijection between the sets of isomorphism classes of principal homogeneous spaces of G and G' respectively. An invariant $a \in \operatorname{Inv}^{d+1}\big(G, \mathbb{Q}/\mathbb{Z}(d)\big)$ defines an invariant $a' \in \operatorname{Inv}^{d+1}\big(G', \mathbb{Q}/\mathbb{Z}(d)\big)$ by the formula

$$a'_E\big(R_{L\otimes E/E}(Y)\big) = \operatorname{Cor}_{L\otimes E/E}\big(a_{L\otimes E}(Y)\big)$$

for a principal homogeneous space Y of G over $L\otimes E$. Hence, we get the *corestriction* homomorphism

$$\operatorname{Cor}_{L/F} : \operatorname{Inv}_L^{d+1}\big(G, \mathbb{Q}/\mathbb{Z}(d)\big) \to \operatorname{Inv}_F^{d+1}\big(G', \mathbb{Q}/\mathbb{Z}(d)\big).$$

Write $\operatorname{Res}_{L/F}$ for the composition

$$\operatorname{Inv}_F^{d+1}\big(G', \mathbb{Q}/\mathbb{Z}(d)\big) \to \operatorname{Inv}_L^{d+1}\big(G'_L, \mathbb{Q}/\mathbb{Z}(d)\big) \to \operatorname{Inv}_L^{d+1}\big(G, \mathbb{Q}/\mathbb{Z}(d)\big),$$

where the first homomorphism is the restriction map and the second one is induced by the natural group homomorphism $G \to G'_L$.

PROPOSITION 9.8. *Let L/F be a finite separable field extension, let G be a simply connected group over L, and set $G' = R_{L/F}(G)$. Then the natural homomorphism*

$$\operatorname{Cor}_{L/F} : \operatorname{Inv}_L^3\big(G, \mathbb{Q}/\mathbb{Z}(2)\big)_{\mathrm{norm}} \to \operatorname{Inv}_F^3\big(G', \mathbb{Q}/\mathbb{Z}(2)\big)_{\mathrm{norm}}$$

is an isomorphism.

PROOF. The composition $\operatorname{Res}_{L/F} \circ \operatorname{Cor}_{L/F}$ is the identity. In particular, $\operatorname{Cor}_{L/F}$ is injective. Surjectivity follows from the surjectivity of r_G and Proposition 7.7. □

Let G be a simply connected group. We have

$$G \simeq \prod_{j=1}^{s} R_{E_j/F}(G_j),$$

where E_j/F is a finite separable field extension and G_j is an absolutely simple group over E_j. Propositions 9.7 and 9.8 yield:

PROPOSITION 9.9. *There is a canonical isomorphism*

$$\operatorname{Inv}^3_F\big(G, \mathbb{Q}/\mathbb{Z}(2)\big)_{\text{norm}} \xrightarrow{\ \sim\ } \textstyle\coprod_{j=1}^{s} \operatorname{Inv}^3_{E_j}\big(G_j, \mathbb{Q}/\mathbb{Z}(2)\big)_{\text{norm}}.$$

Thus, the study of degree 3 cohomological invariants reduces to the case of an absolutely simple group.

Let G be an absolutely simple simply connected group. The group $Q(G)$ is infinite cyclic with canonical generator q_G. The invariant $R_G = r_G(q_G)$ is called the *Rost invariant of* G.

Proposition 9.2 yields:

THEOREM 9.10. *For every principal homogeneous space Y of an absolutely simple simply connected group G, the kernel of the natural homomorphism*

$$H^3\big(F, \mathbb{Q}/\mathbb{Z}(2)\big) \to H^3\big(F(Y), \mathbb{Q}/\mathbb{Z}(2)\big)$$

is a (finite) cyclic group generated by $R_G(Y)$.

THEOREM 9.11. *For an absolutely simple simply connected group G, the group $\operatorname{Inv}^3\big(G, \mathbb{Q}/\mathbb{Z}(2)\big)_{\text{norm}}$ is finite cyclic with canonical generator R_G. For a homomorphism $\alpha : G \to G'$ of simply connected groups, $\alpha^*(R_{G'}) = n_\alpha R_G$, where n_α is the Rost multiplier of α.*

PROOF. The first statement follows from the fact that $Q(G)$ is a cyclic group generated by q_G. The second statement is a consequence of Proposition 9.4 and the definition of the Rost multiplier. $\square$

We will need the following lemma in the next section. Let G be a simply connected group. Recall that $X = S/\rho(G)$ is a classifying variety of G and $\pi : S \to X$ is the canonical morphism. Assume in addition that S is a special simply connected group.

LEMMA 9.12. *Let $x \in X$ be a point, set $Y = \pi^{-1}x$ (so that Y is a principal homogeneous space of G over $F(x)$), write $\varepsilon : Y \to S$ for the natural morphism. Then the composition*

$$A^1(S, K_2) \xrightarrow{\ \varepsilon^*\ } A^1(Y, K_2) \xrightarrow{\ \mu_Y\ } Q(G_{F(x)})$$

coincides with the composition

$$A^1(S, K_2) \xrightarrow{\ \mu_S\ } Q(S) \xrightarrow{\ Q(\rho)\ } Q(G) \xrightarrow{\ \text{Res}\ } Q(G_{F(x)}).$$

PROOF. Extending the base field to $F(x)$ we may assume that $F = F(x)$, i.e., x is a rational point. We also may assume that F is separably closed. Choose an isomorphism $\beta : Y \xrightarrow{\sim} G$. The composition

$$Y \xrightarrow{\ \beta\ } G \xrightarrow{\ \rho\ } S$$

and ε differ by a translation by an element of $S(F)$. Hence, by Lemma 6.9, these morphisms induce the same pullback homomorphisms. The statement follows now from Theorem 9.3 applied to the group homomorphism $\rho : G \to S$. □

10. The order of the Rost invariant

In this section we compute the order of the Rost invariant of a group G in terms of the representation theory of G.

Let G be an absolutely simple simply connected group. By Theorem 9.11, the group $\operatorname{Inv}^3\big(G, \mathbb{Q}/\mathbb{Z}(2)\big)_{\mathrm{norm}}$ is finite cyclic and generated by the Rost invariant R_G. Write n_G for the order of R_G. How can one compute the numbers n_G?

We start with some simple relations between the numbers n_G and the Rost multipliers n_μ defined in Section 7.

LEMMA 10.1. *Let G be an absolutely simple simply connected group.*

(1) *For a homomorphism $\alpha : G \to G'$ of absolutely simple simply connected groups, n_G divides $n_\alpha \cdot n_{G'}$.*
(2) *For every representation $\mu : G \to \mathbf{SL}(V)$, n_G divides n_μ.*

PROOF. (1) By Theorem 9.11, $\alpha^*(R_{G'}) = n_\alpha R_G$, hence $(n_\alpha \cdot n_{G'})R_G = 0$.
(2) The statement follows from (1) with $G' = \mathbf{SL}_n$ since $n_{\mathbf{SL}_n} = 1$. □

Thus, it follows from Lemma 10.1 that n_G divides the number $\gcd n_\mu$ (known as the *Dynkin index* of G, see [Dy 52], [LS 97]), where μ ranges over all representations of G. In this section we prove that $n_G = \gcd n_\mu$.

Choose an injective group homomorphism $\rho : G \hookrightarrow S$, where S is a simply connected special group (for example, we can take $S = \mathbf{SL}_n$ for some n). Set $X = S/\rho(G)$, and write

$$\pi : S \to X$$

for the canonical morphism. Consider the spectral sequence associated with π [Rost 96, Sec. 8]:

$$E_1^{p,q} = \coprod_{x \in X^{(p)}} A^q(\pi^{-1}x, K_{2-p}) \Longrightarrow A^{p+q}(S, K_2).$$

The E_1-term looks as follows:

$$\begin{array}{ccccc} A^1(Y_\xi, K_2) & \longrightarrow & \coprod\limits_{x \in X^{(1)}} A^1(\pi^{-1}x, K_1) & & \\ & & & & \\ A^0(Y_\xi, K_2) & \longrightarrow & \coprod\limits_{x \in X^{(1)}} A^0(\pi^{-1}x, K_1) & \longrightarrow & \coprod\limits_{x \in X^{(2)}} A^0(\pi^{-1}x, K_0), \end{array}$$

where Y_ξ is the generic fiber of π.

By Corollary 5.4,

$$E_1^{1,0} = \coprod_{x \in X^{(1)}} K_1 F(x) \qquad \text{and} \qquad E_1^{2,0} = \coprod_{x \in X^{(1)}} K_0 F(x),$$

hence $E_2^{2,0} = \mathrm{CH}^2(X)$. It follows from Proposition 6.1 that $E_1^{1,1} = 0$ and $A^2(S, K_2) = \mathrm{CH}^2(S) = 0$. Hence, the spectral sequence induces an exact sequence

$$A^1(S, K_2) \xrightarrow{\varepsilon^*} A^1(Y_\xi, K_2) \xrightarrow{d} \mathrm{CH}^2(X) \to 0,$$

where d is the differential $E_2^{0,1} \to E_2^{2,0}$ in the spectral sequence and $\varepsilon : Y_\xi \to S$ is the natural morphism. Lemma 9.12 yields the following commutative diagram:

$$\begin{array}{ccccccc} & & A^1(S, K_2) & \xrightarrow{\varepsilon^*} & A^1(Y_\xi, K_2) & \xrightarrow{d} & \mathrm{CH}_2(X) \longrightarrow 0 \\ & & {\scriptstyle \mu_S}\downarrow\wr & & {\scriptstyle \tau_0}\downarrow\wr & & \\ Q(S) & = & Q(S)_0 & \xrightarrow{Q(\rho)} & Q(G)_0, & & \end{array}$$

where τ_0 is the isomorphism from Corollary 9.6.

Let $e(\rho)$ denote the composition $d \circ \tau_0^{-1}$. We have proved:

PROPOSITION 10.2. *The sequence*

$$Q(S)_0 \xrightarrow{Q(\rho)} Q(G)_0 \xrightarrow{e(\rho)} \mathrm{CH}^2(X) \to 0$$

is exact.

Suppose that we are given another injective group homomorphism $\rho' : G \hookrightarrow S'$ into a special simply connected group S' and a surjective group homomorphism $\alpha : S' \to S$ such that the composition $\alpha \circ \rho'$ coincides with ρ. Let $X' = S'/\rho'(G)$. The homomorphism α induces a surjective morphism $f : X' \to X$.

LEMMA 10.3. *The following diagram commutes:*

$$\begin{array}{ccccccc} Q(S)_0 & \xrightarrow{Q(\rho)} & Q(G)_0 & \xrightarrow{e(\rho)} & \mathrm{CH}^2(X) & \longrightarrow & 0 \\ {\scriptstyle Q(\alpha)}\downarrow & & \| & & \downarrow{\scriptstyle f^*} & & \\ Q(S')_0 & \xrightarrow{Q(\rho')} & Q(G)_0 & \xrightarrow{e(\rho')} & \mathrm{CH}^2(X') & \longrightarrow & 0. \end{array}$$

PROOF. By [Rost 96, Th. 12.7], the spectral sequences associated to π and $\pi' : S' \to X'$ commute with the pullback homomorphisms, i.e., the diagram

$$\begin{array}{ccccc} A^1(S, K_2) & \xrightarrow{i^*} & A^1(Y_\xi, K_2) & \xrightarrow{d} & \mathrm{CH}^2(X) \\ {\scriptstyle \alpha^*}\downarrow & & \downarrow & & \downarrow{\scriptstyle f^*} \\ A^1(S', K_2) & \xrightarrow{i'^*} & A^1(Y'_\xi, K_2) & \xrightarrow{d'} & \mathrm{CH}^2(X') \end{array}$$

is commutative. Since $Y'_{\xi'} \cong (Y_\xi)_{F(X')}$, the statement now follows from Theorem 9.3 and the fact that μ_{Y_ξ} commutes with field extensions. □

Let $\mu : G \to \mathbf{SL}(V)$ be a representation of a simply connected group G. The group $Q(\mathbf{SL}(V))_0$ is infinite cyclic with canonical generator $q_{\mathbf{SL}(V)}$. Denote its image $Q(\mu)(q_{\mathbf{SL}(V)}) \in Q(G)$ by q_μ. Note that if G is absolutely simple, $q_\mu = n_\mu \cdot q_G$.

Let $\rho : G \hookrightarrow \mathbf{SL}_n$ be a faithful representation, set $X = \mathbf{SL}_n / \rho(G)$, and write

$$\pi : \mathbf{SL}_n \to X$$

for the canonical morphism. By Proposition 10.2, $e(\rho)$ induces an isomorphism

$$e : Q(G)_0 / \mathbb{Z} q_\rho \xrightarrow{\sim} \mathrm{CH}^2(X).$$

The group G acts (on the right) on the trivial vector bundle $\mathbf{SL}_n \times V$ over $\mathbf{SL}_n$ by the formula

$$(a, v) \cdot g = \big(a\rho(g), \mu(g^{-1})(v)\big).$$

Write E_μ for the vector bundle $(\mathbf{SL}_n \times V)/G$ over X. We will use the following:

THEOREM 10.4. [Mer 97, Cor. 4.7] *The group* $K_0(X)$ *is generated by the classes* $[E_\mu]$ *as* μ *ranges over all representations of the group* G *over* F.

Let $\operatorname{Rep}(G)$ be the abelian category of finite dimensional representations of G and let $R(G) = K_0\big(\operatorname{Rep}(G)\big)$ denote the Grothendieck ring of the category $\operatorname{Rep}(G)$. Consider the homomorphism

$$b : R(G) \to K_0(X) \quad \text{given by} \quad [\mu] \mapsto [E_\mu]$$

and the composition

$$b_\rho : R(G) \xrightarrow{b} K_0(X) \xrightarrow{c_2} \mathrm{CH}^2(X) \xrightarrow{e^{-1}} Q(G)_0/\mathbb{Z}\, q_\rho,$$

where c_2 is the second Chern class.

LEMMA 10.5. *The map* b_ρ *is a surjective homomorphism.*

PROOF. It follows from the exactness of the sequence [San 81, Prop. 6.10]

$$1 = (\mathbf{SL}_n)^* \to \mathrm{CH}^1(X) \to \mathrm{CH}^1(G) = 0$$

that $\mathrm{CH}^1(X) = 0$. Hence, the the additivity formula $c_2(x+y) = c_2(x) + c_2(y) + c_1(x)c_1(y)$ for the second Chern class implies that c_2 is a homomorphism. The class c_2 is surjective (see the proof of Proposition 6.1). Finally, by Theorem 10.4, b is also surjective. □

PROPOSITION 10.6. *For every simply connected group* G, *the group* $Q(G)_0$ *is generated by the elements* q_μ *as* μ *ranges over the representations* $\mu \in \operatorname{Rep}(G)$.

PROOF. We fix a faithful representation $\rho : G \hookrightarrow \mathbf{SL}_n$ and set $X = \mathbf{SL}_n / \rho(G)$. In view of Lemma 10.5, it is sufficient to prove that for every representation $\mu : G \to \mathbf{SL}(V)$, the image $b_\rho(\mu)$ belongs to the subgroup of $Q(G)_0/\mathbb{Z}\, q_\rho$ generated by $q_\mu + \mathbb{Z}\, q_\rho$.

Consider the factor variety $X' = \big(\mathbf{SL}_n \times \mathbf{SL}(V)\big)/G$ with respect to the diagonal action of G. Let $f : X' \to X$ be the canonical morphism. By Proposition 10.2 applied to the homomorphism $(\rho, \mu) : G \to \mathbf{SL}_n \times \mathbf{SL}(V)$, $e(\rho, \mu)$ induces an isomorphism

$$e' : Q(G)_0/(\mathbb{Z}\, q_\rho + \mathbb{Z}\, q_\mu) \xrightarrow{\sim} \mathrm{CH}^2(X'),$$

and it follows from Lemma 10.3 that the diagram

$$\begin{array}{ccc} Q(G)_0/\mathbb{Z}\, q_\rho & \xrightarrow{e} & \mathrm{CH}^2(X) \\ {\scriptstyle i}\downarrow & & \downarrow{\scriptstyle f^*} \\ Q(G)_0/(\mathbb{Z}\, q_\rho + \mathbb{Z}\, q_\mu) & \xrightarrow{e'} & \mathrm{CH}^2(X'), \end{array}$$

commutes, where i is the canonical surjection.

The morphism $\mathbf{SL}_n \times \mathbf{SL}(V) \times V \to \mathbf{SL}_n \times V$ taking (a, b, v) to $(a, b^{-1}(v))$ gives rise to a morphism $X' \times V \to E_\mu$ which induces a trivialization of the vector bundle $f^* E_\mu$ over X'. Hence the image of $c_2(E_\mu)$ under $f^* : \mathrm{CH}^2(X) \to \mathrm{CH}^2(X')$ is trivial and therefore, $b_\rho(\mu)$ belongs to the kernel of the map i in the diagram, i.e., the class $b_\rho(\mu) + \mathbb{Z}\, q_\rho$ in $Q(G)_0/\mathbb{Z}\, q_\rho$ belongs to the subgroup generated by $q_\mu + \mathbb{Z}\, q_\rho$. □

THEOREM 10.7. *For every absolutely simple simply connected group G,*

$$n_G = \gcd(n_\mu),$$

where μ ranges over all representations of G.

PROOF. Let $n = \gcd n_\mu$. Since $q_\mu = n_\mu \cdot q_G$, by Proposition 10.6 we have $Q(G)_0 = nq_G$. Hence, in view of Proposition 9.5,

$$\operatorname{Inv}^3\big(G, \mathbb{Q}/\mathbb{Z}(2)\big) \cong Q(G)/Q(G)_0 = \mathbb{Z}q_G/n\mathbb{Z}q_G$$

is a cyclic group of order n. The number n_G is the order of the generator R_G of this group, hence $n = n_G$. □

REMARK 10.8. The n-torsion part of $H^3\big(F, \mathbb{Q}/\mathbb{Z}(2)\big)$ is naturally isomorphic to $H^3\big(F, \mathbb{Z}/n_G\mathbb{Z}(2)\big)$, see Appendix A. Hence the Rost invariant R_G takes values in $H^3(F, \mathbb{Z}/n_G\mathbb{Z}(2))$ and n_G is the smallest natural number with this property.

10.9. Let G be a simply connected semisimple group over F, let $\mu : G \to \mathbf{SL}(V)$ be a representation over F, and fix $T \subset G$ a maximal torus defined over F. The space $V_{\text{sep}} = V \otimes_F F_{\text{sep}}$ splits into a direct sum of one-dimensional eigenspaces with some eigencharacters $\chi_1, \chi_2, \dots, \chi_m \in T^*_{\text{sep}}$. Then, by Example 6.15 and Proposition 7.9(2),

$$n_\mu = \frac{1}{2} \sum_i \langle \chi_i, \eta_G \rangle^2 \in \mathbb{Z},$$

where $\eta_G \in T_{\text{sep}*}$ is a short coroot of G.

Consider the *character* (restriction) homomorphism

$$\operatorname{ch}_G : R(G_{\text{sep}}) \to R(T_{\text{sep}}) = \mathbb{Z}[T^*_{\text{sep}}]$$

and the homomorphism

$$N_G : \mathbb{Z}[T^*_{\text{sep}}] \to \frac{1}{2}\,\mathbb{Z},$$

which takes a character χ to $\frac{1}{2}\langle \chi, \eta_G \rangle^2$, so that for a representation μ of G,

$$(N_G \circ \operatorname{ch}_G)([\mu]) = n_\mu.$$

Thus, by Theorem 10.7,

$$(N_G \circ \operatorname{ch}_G)(R(G)) = (\gcd n_\mu)\mathbb{Z} = n_G\mathbb{Z}. \tag{10.10}$$

The Weyl group W of G_{sep} acts naturally on $R(T_{\text{sep}})$. Fix a system of simple roots. The absolute Galois group $\Gamma = \operatorname{Gal}(F)$ acts on $R(T_{\text{sep}})$ through the $*$-action on T^*_{sep} defined in [Ti 66, 2.3]. Therefore, the semidirect product Δ of W and Γ acts naturally on $R(T_{\text{sep}}) = \mathbb{Z}[T^*_{\text{sep}}]$.

Write $\Lambda^+ \subset T^*_{\text{sep}}$ for the cone of dominant characters. The group Γ leaves Λ^+ invariant. The field corresponding to the stabilizer $\Gamma_\rho \subset \Gamma$ of a character $\rho \in \Lambda^+$ by Galois theory is called the *field of definition* of ρ; we denote it by F_ρ.

Let C denote the center of G. For a character $\rho \in T^*_{\text{sep}}$, write $\bar\rho \in C^*_{\text{sep}}$ for its restriction to C_{sep}. The field $F_{\bar\rho}$ corresponding to the stabilizer $\Gamma_{\bar\rho} \subset \Gamma$ of $\bar\rho$ is called the *field of definition* of $\bar\rho$. For a dominant character $\rho \in \Lambda^+$, the field of definition $F_{\bar\rho}$ of $\bar\rho$ is contained in F_ρ. Denote by $A_{\bar\rho}$ a *Tits algebra* associated to $\bar\rho$ as in [KMRT 98, §27] or [Ti 71, §4], so that $A_{\bar\rho}$ is a central simple algebra over $F_{\bar\rho}$

uniquely determined up to Brauer equivalence over $F_{\bar\rho}$. For every character $\rho \in \Lambda^+$ set

$$A_\rho = A_{\bar\rho} \otimes_{F_{\bar\rho}} F(\rho).$$

The algebra A_ρ is a central simple over F_ρ. The index of A_ρ depends only on the Γ-orbit of ρ.

For a dominant character $\rho \in \Lambda^+$, let $\Delta\rho$, $W\rho$, $\Gamma\rho$ denote the corresponding orbits. We write $\Delta(\rho) \in \mathbb{Z}[T^*_{\mathrm{sep}}]$ for the sum in $R(T_{\mathrm{sep}}) = \mathbb{Z}[T^*_{\mathrm{sep}}]$ of all (finitely many) characters in $\Delta\rho$. We define $W(\rho), \Gamma(\rho) \in \mathbb{Z}[T^*_{\mathrm{sep}}]$ similarly.

Let μ and ρ be characters in T^*_{sep}. We write $\rho > \mu$ if $\rho - \mu$ is a sum of simple roots. This gives a partial ordering on T^*_{sep}. Clearly, if $\rho > \mu$, then $\bar\rho = \bar\mu$ and the Tits algebras $A_{\bar\rho}$ and $A_{\bar\mu}$ are Brauer equivalent.

THEOREM 10.11. *The elements* $\operatorname{ind}(A_\rho) \cdot \Delta(\rho)$ *as* ρ *varies over* Λ^+ *generate the subgroup* $\operatorname{ch}_G\bigl(R(G)\bigr) \subset \mathbb{Z}[T^*_{\mathrm{sep}}]$.

PROOF. Let ρ be a dominant character and let V_ρ be the irreducible representation of G with the highest character ρ (defined over F_{sep}). By the representation theory of semisimple groups [Hu 75],

$$\operatorname{ch}_G(V_\rho) = \sum_{\mu \le \rho} c_\mu \, W(\mu)$$

for some $c_\mu \in \mathbb{Z}$, $\mu \in \Lambda^+$, with $c_\rho = 1$. The representation theory over arbitrary fields [Ti 71] shows that the representation $V_\rho^{\oplus \operatorname{ind} A_\rho}$ is defined over the field F_ρ, hence the character $\operatorname{ch}_G(V_\rho)$ is Γ_ρ-invariant. It follows that $c_\mu = c_{\mu'}$ for every two dominant characters $\mu < \rho$ and $\mu' < \rho$ such that $\Gamma_\rho\mu = \Gamma_\rho\mu'$. Let I be a set of representatives of Γ_ρ-orbits in the set of dominant characters $\mu < \rho$. We have

$$\operatorname{ch}_G(V_\rho) = W(\rho) + \sum_{\mu \in I} c_\mu \, W\Gamma_\rho(\mu)$$

Let J be a complete set of coset representatives of Γ_ρ in Γ. The representation

$$W_\rho = \coprod_{\gamma \in J} (V_{\gamma\rho})^{\oplus \operatorname{ind} A_\rho}$$

is defined over F and the classes of these W_ρ generate $R(G)$ as ρ ranges over Λ^+ [Ti 71, Th. 3.3]. We have

$$\sum_{\gamma \in J} \gamma\Gamma_\rho(\mu) = [\Gamma_\mu : \Gamma_\rho \cap \Gamma_\mu] \cdot \Gamma(\mu),$$

hence

$$\sum_{\gamma \in J} \gamma W\Gamma_\rho(\mu) = [\Gamma_\mu : \Gamma_\rho \cap \Gamma_\mu] \cdot \Delta(\mu),$$

and

$$\begin{aligned}\operatorname{ch}_G(W_\rho) &= \sum_{\gamma\in J} \operatorname{ind} A_\rho \cdot \gamma W(\rho) + \sum_{\gamma\in J}\sum_{\mu\in I} c_\mu \operatorname{ind} A_\rho \cdot \gamma W \Gamma_\rho(\mu) \\ &= \operatorname{ind}(A_\rho)\cdot \Delta(\rho) + \sum_{\gamma\in J}\sum_{\mu\in I} c_\mu \operatorname{ind} A_\rho \cdot [\Gamma_\mu : \Gamma_\rho \cap \Gamma_\mu] \cdot \Delta(\mu) \\ &= \operatorname{ind}(A_\rho)\cdot \Delta(\rho) + \sum_{\gamma\in J}\sum_{\mu\in I} c_\mu b_\mu \cdot \big(\operatorname{ind} A_\mu \cdot \Delta(\mu)\big),\end{aligned}$$

where

$$b_\mu = \frac{\operatorname{ind} A_\rho \cdot [\Gamma_\mu : \Gamma_\rho \cap \Gamma_\mu]}{\operatorname{ind} A_\mu}.$$

Let L be the extension of F corresponding to $\Gamma_\rho \cap \Gamma_\mu$ by Galois theory, so that $F_\rho \subset L$ and $F_\mu \subset L$. Since $\mu < \rho$, the algebras $A_\rho \otimes_{F_\rho} L$ and $A_\mu \otimes_{F_\mu} L$ are Brauer equivalent. We have

$$\operatorname{ind} A_\mu \mid \operatorname{ind}(A_\mu \otimes_{F_\mu} L)\cdot[L : F_\mu] = \operatorname{ind}(A_\rho \otimes_{F_\rho} L)\cdot[\Gamma_\mu : \Gamma_\rho \cap \Gamma_\mu] \mid \operatorname{ind} A_\rho \cdot [\Gamma_\mu : \Gamma_\rho \cap \Gamma_\mu],$$

therefore, $b_\mu \in \mathbb{Z}$.

The statement of the theorem now follows by induction on the partial ordering $<$ of Λ^+. □

The formula (10.10) then yields:

COROLLARY 10.12. *For an absolutely simple simply connected group G,*

$$n_G = \gcd_{\rho\in\Lambda^+} \big[\operatorname{ind}(A_\rho)\cdot N_G\big(\Delta(\rho)\big)\big].$$

REMARK 10.13. Corollary 10.12 shows that the number n_G depends only on the Galois action on the Dynkin diagram of G and indices of Tits algebras of G. In other words, if two groups G and G' are strongly inner forms of each other (see [Ti 90]), then $n_G = n_{G'}$.

10.14. THE SPLIT CASE. Now assume that G is a split simply connected group. Let $\rho_1, \rho_2, \ldots, \rho_r$ be the fundamental representations of G, and write $\chi_1, \chi_2, \ldots, \chi_r \in \Lambda^+$ respectively for their highest characters. The character map ch_G identifies the ring $R(G)$ with the polynomial ring [Sam 90, §3.5]

$$\mathbb{Z}[\rho_1, \rho_2, \ldots, \rho_r] = \mathbb{Z}[\Delta(\chi_1), \Delta(\chi_2), \ldots, \Delta(\chi_r)].$$

By Proposition 7.9.5b, the map $N_G \circ \operatorname{ch}_G$ is a derivation, i.e., it satisfies

$$(N_G \circ \operatorname{ch}_G)(xy) = (N_G \circ \operatorname{ch}_G)(x)\cdot \operatorname{rank}(y) + (N_G \circ \operatorname{ch}_G)(y)\cdot \operatorname{rank}(x),$$

where the rank map takes the class of a representation ρ to $\dim(\rho)$. Hence,

$$n_G = \gcd_i n_{\rho_i} = \gcd_i N_G\big(\Delta(\chi_i)\big). \tag{10.15}$$

The numbers n_{ρ_i} may depend on the characteristic, but their gcd does not since the second gcd in (10.15) is characteristic-free. The Dynkin indices n_G are given in [LS 97, Prop. 2.6]:

type of G	A_n	B_n $(n \geq 3)$	C_n	D_n $(n \geq 4)$	E_6	E_7	E_8	F_4	G_2
n_G	1	2	1	2	6	12	60	6	2

(10.16) DYNKIN INDEX FOR G SPLIT

11. Dynkin index for groups of type ^1A

We compute the number n_G for an absolutely simple simply connected group G of type $^1A_{n-1}$ over a field F with $n \geq 2$. Let $T \subset G$ be a maximal torus defined over F. The group of characters $T^*_{\rm sep}$ can be identified with $\mathbb{Z}^n/\mathbb{Z}$ (with $\mathbb{Z}$ embedded diagonally) and $T_{{\rm sep}*}$ with the subgroup of $\mathbb{Z}^n$ of elements whose components sum is equal to zero.

Let e_1, e_2, ..., e_n be the standard basis of $\mathbb{Z}^n$. We write $\bar{e}_i$ for the image of e_i in $\mathbb{Z}^n/\mathbb{Z}$. All the coroots $\pm(e_i - e_j)$ have the same length and we can take $\eta_G = e_1 - e_2$. The Weyl group is the symmetric group S_n which permutes the $\bar{e}_i$ (see Example 6.15). The restriction homomorphism to the center C of G

$$\mathbb{Z}^n/\mathbb{Z} = T^*_{\rm sep} \to C^*_{\rm sep} = \mathbb{Z}/n\mathbb{Z},$$

takes $\sum x_i\bar{e}_i$ to $\sum x_i + n\mathbb{Z}$.

We choose the set of simple roots $\{\bar{e}_1 - \bar{e}_2, \bar{e}_2 - \bar{e}_3, \dots, \bar{e}_{n-1} - \bar{e}_n\}$. The corresponding cone of dominant characters Λ^+ consists of all $\sum x_i\bar{e}_i$ such that $x_1 \geq x_2 \geq \cdots \geq x_n$, see e.g. [Bou 81] or [GW 00].

Choose a character $\chi \in T^*_{\rm sep}$. Some of the coordinates of χ may coincide. Let χ have distinct coordinates (in some order) $a_1 > a_2 > \cdots > a_k$ which repeat $r_1, r_2, \dots, r_k$ times respectively, so that $n = \sum r_i$. Note that the a_i can be modified by adding an integer to all the a_i. We denote the character χ by $(r_1, \dots, r_k; a_1, \dots, a_k)$ or simply by $(\mathbf{r}, \mathbf{a})$.

The stabilizer H of $\chi = (\mathbf{r}, \mathbf{a})$ in the Weyl group S_n is isomorphic to $S_{r_1} \times S_{r_2} \times \cdots \times S_{r_k}$. Hence the number of characters in the S_n-orbit of χ is equal to

$$\frac{n!}{r_1!\,r_2!\cdots r_k!}.$$

For a pair of distinct indices (i, j) the number of characters in the S_n-orbit with first two coordinates (a_i, a_j) is equal to

$$\frac{(n-2)!\; r_i r_j}{r_1!\,r_2!\cdots r_k!}.$$

For such characters χ' we have $\langle \chi', \eta_G \rangle = a_i - a_j$. Denote by $S_n(\chi)$ the sum of characters in $R(T_{\rm sep})$ of the S_n-orbit of χ. We have the following computation:

$$\begin{aligned}
N_G\bigl(S_n(\chi)\bigr) &= \frac{1}{2} \sum_{w \in S_n/H} \langle w\chi, \eta_G \rangle^2 \\
&= \frac{1}{2} \cdot \frac{(n-2)!}{r_1!\,r_2!\cdots r_k!} \sum_{(i,j)} r_i r_j (a_i - a_j)^2 \\
&= \frac{(n-2)!}{r_1!\,r_2!\cdots r_k!} \left[\Bigl(\sum_i r_i\Bigr)\Bigl(\sum_i r_i a_i^2\Bigr) - \Bigl(\sum_i r_i a_i\Bigr)^2 \right] \\
&= \frac{(n-2)!}{r_1!\,r_2!\cdots r_k!} \left[n\Bigl(\sum_i r_i a_i^2\Bigr) - \Bigl(\sum_i r_i a_i\Bigr)^2 \right].
\end{aligned}$$

Denote this integer by $[r_1, \dots, r_k; a_1, \dots, a_k]$ or simply by $[\mathbf{r}, \mathbf{a}]$. We also set $\mathbf{ra} = \sum r_i a_i$.

Let G be a simply connected group of inner type A_{n-1}, i.e., $G = \mathbf{SL}_1(A)$, where A is a central simple algebra of degree $n \geq 2$ over F. We have $\Delta = W = S_n$. The Tits algebra of a character $(\mathbf{r}, \mathbf{a})$ is similar to $A^{\otimes \mathbf{ra}}$ by [KMRT 98, §27.B]. Hence, in view of Corollary 10.12,

$$n_G = \gcd\bigl(\operatorname{ind}(A^{\otimes \mathbf{ra}}) \cdot [\mathbf{r}, \mathbf{a}]\bigr), \tag{11.1}$$

where the gcd is taken over all characters $(\mathbf{r}, \mathbf{a})$ such that $\sum r_i = n$.

Write v_p for the p-adic valuation on $\mathbb{Z}$. For any integer $c \geq 0$, let $s_p(c)$ be the sum of the digits in the base p expansion of c.

LEMMA 11.2. *If $c = c_1 + c_2 + \cdots + c_k$, $c_i \geq 0$, then*

$$v_p\left(\frac{c!}{c_1!\, c_2! \cdots c_k!}\right) = \frac{\sum s_p(c_i) - s_p(c)}{p-1}.$$

PROOF. This is a direct consequence of the formula

$$v_p(n!) = \frac{n - s_p(n)}{p-1}$$

which can be found in, e.g., [Ne 99, II.5.6]. □

LEMMA 11.3. *Let p be a prime integer, and let $n = r_1 + r_2 + \cdots + r_k$ with $r_i \geq 1$ for all i. Write $\ell = \min v_p(r_i)$ and let j be such that $v_p(r_j) = \ell$. Then*

$$v_p\left(\frac{n!}{r_1!\, r_2! \cdots r_k!}\right) \geq v_p(n) - \ell,$$

and the equality holds if and only if

$$s_p(n-1) = s_p(r_1) + \cdots + s_p(r_j - 1) + \cdots + s_p(r_k).$$

PROOF. We have

$$\frac{n!}{r_1!\, r_2! \cdots r_k!} = \frac{n}{r_j} \cdot \frac{(n-1)!}{r_1! \cdots (r_j - 1)! \cdots r_k!}$$

and the second factor of the r.h.s. is integral, whence the inequality. The second statement follows from Lemma 11.2 applied to the second factor. □

LEMMA 11.4. *For every dominant character $(\mathbf{r}, \mathbf{a})$, $\gcd(n, \mathbf{ra})$ divides $[\mathbf{r}, \mathbf{a}]$.*

PROOF. Let p be a prime divisor of n and set $\ell = \min v_p(r_i)$. Obviously,

$$v_p\left(n \cdot \sum_i r_i a_i^2\right) \geq v_p(n) + \ell \quad \text{and} \quad v_p(\mathbf{ra}^2) \geq v_p(\mathbf{ra}) + \ell.$$

By Lemma 11.3,

$$v_p\left(\frac{(n-2)!}{r_1! r_2! \cdots r_k!}\right) \geq -\ell.$$

Hence,

$$v_p([\mathbf{r}, \mathbf{a}]) \geq -\ell + \min\bigl(v_p(n) + \ell, v_p(\mathbf{ra}) + \ell\bigr) = \min\bigl(v_p(n), v_p(\mathbf{ra})\bigr).$$ □

THEOREM 11.5. *Let $G = \mathbf{SL}_1(A)$, where A is a central simple algebra of degree $n \geq 2$ over a field F. Then $n_G = \exp(A)$.*

PROOF. We prove first that n_G divides $\exp(A)$. In view of (11.1), it suffices to show that for every prime number p there is a dominant character $(\mathbf{r},\mathbf{a})$ such that

$$v_p\big(\operatorname{ind}(A^{\otimes \mathbf{ra}})\cdot[\mathbf{r},\mathbf{a}]\big)=v_p\big(\exp(A)\big).$$

Let $a=v_p\big(\exp(A)\big)$. We have $v_p(n)\geq a$ since $\exp(A)\mid n$.

Suppose that $v_p(n)>a$. Consider the character $(\mathbf{r},\mathbf{a})=(p^a,n-p^a;1,0)$. Then

$$[\mathbf{r},\mathbf{a}]=\binom{n-2}{p^a-1},\quad \mathbf{ra}=p^a,\quad\text{and}\quad v_p\big(\operatorname{ind}(A^{\otimes \mathbf{ra}})\big)=0.$$

Clearly, $s_p(n-2)=s_p(n-p^a-1)$. Hence, by Lemma 11.2,

$$v_p\big(\operatorname{ind}(A^{\otimes \mathbf{ra}})\cdot[\mathbf{r},\mathbf{a}]\big)=\frac{s_p(p^a-1)}{p-1}=a=v_p\big(\exp(A)\big).$$

Suppose now that $v_p(n)=a$. Since $\exp(A)\mid\operatorname{ind}(A)\mid n$, it follows that $v_p\big(\operatorname{ind}(A)\big)=a$. Consider the character $(\mathbf{r},\mathbf{a})=(1,n-1;1,0)$. We have $[\mathbf{r},\mathbf{a}]=1=\mathbf{ra}$ and

$$v_p\big(\operatorname{ind}(A^{\otimes \mathbf{ra}})\cdot[\mathbf{r},\mathbf{a}]\big)=v_p\big(\operatorname{ind}(A)\big)=a=v_p\big(\exp(A)\big).$$

It remains to prove that $\exp(A)$ divides $[\mathbf{r},\mathbf{a}]\cdot\operatorname{ind}(A^{\otimes \mathbf{ra}})$ for every dominant character $(\mathbf{r},\mathbf{a})$. By Lemma 11.4,

$$\exp\big(A^{\otimes[\mathbf{r},\mathbf{a}]}\big)\mid\exp\big(A^{\otimes\gcd(n,\mathbf{ra})}\big)=\exp\big(A^{\otimes \mathbf{ra}}\big),$$

and hence

$$\exp(A)\mid\exp(A^{\otimes[\mathbf{r},\mathbf{a}]})\cdot[\mathbf{r},\mathbf{a}]\mid\operatorname{ind}(A^{\otimes \mathbf{ra}})\cdot[\mathbf{r},\mathbf{a}].\qquad\square$$

By Example 2.2, $H^1(F,G)=F^\times/\operatorname{Nrd}(A^\times)$. Consider the normalized invariant r'_G of G defined by

$$r'_G\big(a\operatorname{Nrd}(A^\times)\big)=\{a\}\cup[A],$$

where $[A]$ is the class of the algebra A in the Brauer group $\operatorname{Br}(F)=H^2\big(F,\mathbb{Q}/\mathbb{Z}(1)\big)$. The residue of

$$r'_G\big(t\operatorname{Nrd}(A^\times_{F((t))})\big)=\{t\}\cup[A_{F((t))}]$$

is $[A]$ by formula A.8. On the other hand, $\exp(A)\cdot r'_G=0$. Hence the order of r'_G is equal to $\exp(A)$. It follows from Theorem 11.5 that the invariants r_G and r'_G are two generators of $\operatorname{Inv}^3(G,\mathbb{Q}/\mathbb{Z}(2))_{\text{norm}}$. In particular, any value of the Rost invariant r_G over a field L is the cup product $\{x\}\cup[A_L]$ for some $x\in L^\times$. It seems plausible that r_G coincides with r'_G (up to sign).

12. Dynkin index for groups of type ^2A

Let G be a simply connected group of outer type A_{n-1}, i.e., $G=\mathbf{SU}(B,\tau)$, where B is a central simple algebra of degree $n\geq 3$ with a unitary involution τ over a separable quadratic field extension K/F [KMRT 98, Th. 26.9]. Over the quadratic extension K/F the group G is isomorphic to $\mathbf{SL}_1(B)$. Under the field extension map the Rost invariant r_G maps to the Rost invariant r_{G_K} of order $\exp(B)$ by Theorem 11.5. The corestriction map for the field extension K/F takes r_{G_K} to $2r_G$. Hence

$$\exp(B)\mid n_G\mid 2\exp(B).\tag{12.1}$$

The problem is to decide whether $n_G = \exp(B)$ or $n_G = 2\exp(B)$. Thus, it is sufficient to trace only the 2-part of these integers.

The Galois group $\mathrm{Gal}(F)$ acts on T^*_{sep} through $\mathrm{Gal}(K/F)$ by the involution

$$\kappa\Big(\sum x_i \bar{e}_i\Big) = -\sum x_i \bar{e}_{n+1-i}.$$

A dominant character $\chi = (\mathbf{r}, \mathbf{a}) \in \Lambda^+$ is called *symmetric* if it does not change under κ, that is, the sequence $(r_1, r_2, \dots, r_k)$ is symmetric and the sum $a_i + a_{k+1-i}$ does not depend on i. If $\chi = (\mathbf{r}, \mathbf{a})$ is symmetric, then

$$2\mathbf{ra} = 2\sum_{i=1}^{k} r_i a_i = \sum_{i=1}^{k} r_i(a_i + a_{k+1-i}) \tag{12.2}$$

is divisible by $n = \sum_i r_i$

Let $\chi \in \Lambda^+$ be a dominant character. If χ is symmetric, then $\Delta(\chi) = W(\chi)$. Otherwise, the Δ-orbit of χ is twice as large as the W-orbit of χ. It is also clear that if $(\mathbf{r}', \mathbf{a}') = \kappa(\mathbf{r}, \mathbf{a})$, then $[\mathbf{r}', \mathbf{a}'] = [\mathbf{r}, \mathbf{a}]$.

n is odd

It follows from (12.2) that for a symmetric character $(\mathbf{r}, \mathbf{a})$, $\mathbf{ra}$ is divisible by n. Hence the corresponding Tits algebra A_χ is trivial. If $\chi = (\mathbf{r}, \mathbf{a})$ is not symmetric, then the Tits algebra A_χ is equivalent to $B^{\otimes \mathbf{ra}}$ by [KMRT 98, §27.B].

It follows from Corollary 10.12 that n_G is the gcd of two integers n'_G and n''_G:

$$n'_G = 2\gcd\big(\mathrm{ind}(B^{\otimes \mathbf{ra}}) \cdot [\mathbf{r}, \mathbf{a}]\big),$$

where the gcd is taken over all asymmetric dominant characters $(\mathbf{r}, \mathbf{a})$ and

$$n''_G = \gcd([\mathbf{r}, \mathbf{a}]),$$

where the gcd is taken over all symmetric characters $(\mathbf{r}, \mathbf{a})$.

We can modify the integer n'_G by including in the gcd symmetric characters, without changing the gcd of n'_G and n''_G. It follows from (11.1) and Theorem 11.5 (applied to the algebra B instead of A) that $n'_G = 2\exp(B)$. We get

$$n_G = \gcd\big(2\exp(B),\ \gcd([\mathbf{r}, \mathbf{a}])\big),$$

where the gcd inside the parenthesis is taken over all symmetric characters $(\mathbf{r}, \mathbf{a})$.

We claim that $n_G = 2\exp(B)$, i.e., $[\mathbf{r}, \mathbf{a}]$ is divisible by $2\exp(B)$. By Lemma 11.4, $[\mathbf{r}, \mathbf{a}]$ is divisible by n. It remains to prove that $[\mathbf{r}, \mathbf{a}]$ is even for a symmetric character $(\mathbf{r}, \mathbf{a})$. We have by definition

$$[\mathbf{r}, \mathbf{a}] = \frac{(n-2)!}{r_1!\, r_2! \cdots r_k!} \sum_{i<j} r_i r_j (a_i - a_j)^2.$$

Note that the summands corresponding to the pairs of indices (i, j) and $(k+1-j, k+1-i)$ are equal. Hence, modulo 2, $[\mathbf{r}, \mathbf{a}]$ is congruent to

$$\frac{(n-2)!}{r_1!\, r_2! \cdots r_k!} \sum_{i<k/2} r_i r_{k+1-i}(a_i + a_{k+1-i})^2.$$

Since $n = \sum r_i$ is odd and the character $(\mathbf{r}, \mathbf{a})$ is symmetric, k is odd and therefore, $(a_i + a_{k+1-i}) = 2a_{\frac{k+1}{2}}$ is even. Thus, $[\mathbf{r}, \mathbf{a}]$ is even.

n **is even**

If $\chi = (\mathbf{r}, \mathbf{a})$ is symmetric, then it follows from (12.2) that $\mathbf{ra}$ is divisible by $m = \frac{n}{2}$. The corresponding Tits algebra A_χ is equivalent to $D^{\otimes \frac{\mathbf{ra}}{m}}$, where $D = D(B, \tau)$ is the discriminant algebra of (B, τ) [KMRT 98, §10]. If $\chi = (\mathbf{r}, \mathbf{a})$ is not symmetric, then the Tits algebra A_χ is equivalent to $B^{\otimes \mathbf{ra}}$ by [KMRT 98, §27.B].

It follows from Corollary 10.12 that n_G is the gcd of two integers n'_G and n''_G:

$$n'_G = 2\gcd\big(\operatorname{ind}(B^{\otimes \mathbf{ra}}) \cdot [\mathbf{r}, \mathbf{a}]\big),$$

where the gcd is taken over all asymmetric dominant characters $(\mathbf{r}, \mathbf{a})$ and

$$n''_G = \gcd\big(\operatorname{ind}(D^{\otimes \frac{\mathbf{ra}}{m}}) \cdot [\mathbf{r}, \mathbf{a}]\big),$$

where the gcd is taken over all symmetric characters $(\mathbf{r}, \mathbf{a})$.

Note that the algebra $D^{\otimes \frac{\mathbf{ra}}{m}} \otimes_F K$ is similar to $B^{\otimes \mathbf{ra}}$ [KMRT 98, Prop. 10.30], hence

$$\operatorname{ind}(D^{\otimes \frac{\mathbf{ra}}{m}}) \mid 2\operatorname{ind}(B^{\otimes \mathbf{ra}}).$$

Therefore, we can modify the integer n'_G by including in the gcd also symmetric characters, without changing the gcd of n'_G and n''_G. It follows from (11.1) and Theorem 11.5 (applied to the algebra B instead of A) that $n'_G = 2\exp(B)$. We get

$$n_G = \gcd\left[2\exp(B),\ \gcd\big(\operatorname{ind}(D^{\otimes \frac{\mathbf{ra}}{m}}) \cdot [\mathbf{r}, \mathbf{a}]\big)\right],$$

where the gcd inside the brackets is taken over all symmetric characters $(\mathbf{r}, \mathbf{a})$. Finally,

$$n_G = \begin{cases} 2\exp(B) & \text{if } 2\exp(B) \mid \operatorname{ind}(D^{\otimes \frac{\mathbf{ra}}{m}}) \cdot [\mathbf{r}, \mathbf{a}] \text{ for all symmetric } (\mathbf{r}, \mathbf{a}), \\ \exp(B) & \text{otherwise.} \end{cases}$$

Thus, we need to consider divisibility properties of the integers $[\mathbf{r}, \mathbf{a}] \cdot \operatorname{ind}(D^{\otimes \frac{\mathbf{ra}}{m}})$ for all symmetric characters $(\mathbf{r}, \mathbf{a})$. We only need to look at the 2-part of these integers.

Let $\chi = (\mathbf{r}, \mathbf{a})$ be a symmetric character. We consider two cases.

Suppose first that the integer $\frac{\mathbf{ra}}{m}$ is even, i.e., $\mathbf{ra}$ is divisible by n. We will show (Lemma 12.3) that $v_2([\mathbf{r}, \mathbf{a}]) \geq v_2(2n) \geq v_2(2\exp(B))$, i.e., the term $\operatorname{ind}(D^{\otimes \frac{\mathbf{ra}}{m}}) \cdot [\mathbf{r}, \mathbf{a}]$ does not contribute to the gcd.

LEMMA 12.3. *If n is even and $\mathbf{ra}$ is divisible by n, then $v_2([\mathbf{r}, \mathbf{a}]) \geq v_2(n) + 1$.*

PROOF. Let $\mathbf{ra} = nq$ for some q. We have

$$[\mathbf{r}, \mathbf{a}] = \frac{1}{n-1} \cdot \frac{n!}{r_1!\, r_2! \cdots r_k!} \cdot \left(\sum r_i a_i^2 - nq^2\right).$$

Since $n - 1$ is odd, by Lemma 11.3 it suffices to prove that

$$v_2\left(\sum r_i a_i^2 - nq^2\right) \geq \ell + 1,$$

where $\ell = \min v_2(r_i)$. We have

$$\sum r_i a_i^2 - nq^2 = \sum r_i a_i(a_i + 1) - nq(q + 1).$$

As $v_2(r_i) \geq \ell$ and $v_2(n) \geq \ell$, the r.h.s. is divisible by $2^{\ell+1}$. □

Suppose now that the integer $\frac{\mathbf{ra}}{m}$ is odd.

LEMMA 12.4. *If n is even and $\mathbf{ra} = mq$ with q odd, then $v_2([\mathbf{r}, \mathbf{a}]) \geq v_2(m)$. Equality holds if and only if n is a 2-power, $k = 2$, and $\mathbf{r} = (m, m)$.*

PROOF. By Lemma 11.3,

$$v_2\left(\frac{(n-2)!}{r_1!\, r_2! \cdots r_k!}\right) \geq -\ell, \tag{12.5}$$

where $\ell = \min v_2(r_i)$. Since q is odd, it follows that $v_2(m) = v_2(\mathbf{ra}) \geq \ell$. Thus, to prove the inequality it is sufficient to show that

$$v_2\left(n \sum r_i a_i^2 - m^2 q^2\right) \geq v_2(m) + \ell.$$

This is obvious since $v_2(r_i) \geq \ell$ and $v_2(m) \geq \ell$.

If n is a 2-power, $k = 2$, $\mathbf{r} = (m, m)$, and $\mathbf{a} = (a_1, a_2)$, then

$$\mathbf{ra} = ma_1 + ma_2 = m(a_1 + a_2) = mq,$$

hence $a_1 + a_2$ is odd. We have

$$[\mathbf{r}, \mathbf{a}] = \frac{(2m-2)!}{(m-1)!^2}(a_1 - a_2)^2$$

and by Lemma 11.2, since $a_1 - a_2$ is odd,

$$v_2([\mathbf{r}, \mathbf{a}]) = v_2\left(\frac{(2m-2)!}{(m-1)!^2}\right) = 2s_2(m-1) - s_2(2m-2) = s_2(m-1).$$

Since m is a 2-power, $s_2(m-1) = v_2(m)$, and equality holds in the statemetn of the lemma.

Conversely, assume the equality. Then the first part of the proof shows that the equality in (12.5) holds, or equivalently,

$$v_2\left(\frac{n!}{r_1!\, r_2! \cdots r_k!}\right) = v_2(n) - \ell,$$

and hence by Lemma 11.3,

$$s_2(n-1) = s_2(r_1) + \cdots + s_2(r_j - 1) + \cdots + s_2(r_k),$$

where j satisfies $v(r_j) = \ell$. This means that when we consequently add in base 2 (in any order) the integers $r_1, \ldots, r_j - 1, \ldots, r_k$ we never carry over units. In particular, all these integers are pairwise distinct. Thus, the sequence $\mathbf{r}$ can have at most one pair of equal terms. But the character $(\mathbf{r}, \mathbf{a})$ is symmetric, hence the sequence $\mathbf{r}$ is symmetric. It follows that $k \leq 3$. If $k = 2$, then $\mathbf{r} = (m, m)$ and $s_2(2m-1) = s_2(m) + s_2(m-1)$, i.e., when we add m and $m-1$ (in base 2) we don't carry over units. This is possible only if m (and hence n) is a 2-power.

Finally, assume $k = 3$, i.e., $\mathbf{r} = (r_1, r_2, r_3)$ with $r_1 = r_3$. Then by symmetry $a_1 + a_3 = 2a_2$, hence

$$mq = r_1 a_1 + r_2 a_2 + r_3 a_3 = r_1(a_1 + a_3) + r_2 a_2 = (2r_1 + r_2)a_2 = na_2.$$

Therefore, $q = 2a_2$, which is a contradiction since q is odd. This case does not occur. □

THEOREM 12.6. *Let $G = \mathbf{SU}(B, \tau)$, where B is a central simple algebra of degree $n \geq 3$ with a unitary involution τ over a separable quadratic field extension K/F. Let $D = D(B, \tau)$ be the discriminant algebra of (B, τ). Then*

$$n_G = \begin{cases} \exp(B) & \text{if } n \text{ is a 2-power and } \exp(B) = n; \\ \exp(B) & \text{if } n \text{ is a 2-power, } \exp(B) = \frac{n}{2}, \text{ and } D \text{ is split;} \\ 2\exp(B) & \text{otherwise.} \end{cases}$$

PROOF. If n is odd, we have shown that $n_G = 2\exp(B)$. We may assume that n is even. We know from the Lemmas 12.3 and 12.4 above that $n_G = \exp(B)$ if and only if there exists a symmetric character $(\mathbf{r}, \mathbf{a})$ such that $\mathbf{ra} = mq$ with q odd and

$$v_2\big(\mathrm{ind}(D) \cdot [\mathbf{r}, \mathbf{a}]\big) = v_2\big(\exp(B)\big). \tag{12.7}$$

By Lemma 12.4, for such a character $(\mathbf{r}, \mathbf{a})$ we have

$$v_2([\mathbf{r}, \mathbf{a}]) \geq v_2(m), \tag{12.8}$$

hence the equality (12.7) implies

$$v_2(m) + 1 = v_2(n) \geq v_2\big(\exp(B)\big) \geq v_2\big(\mathrm{ind}(D)\big) + v_2(m) \geq v_2(m).$$

There are two cases:

Case 1: $v_2\big(\exp(B)\big) = v_2(n) = v_2(m) + 1$.

The algebra D_K is similar to $B^{\otimes m}$ and hence is not split. The index $\mathrm{ind}(D)$ divides 4 [KMRT 98, Prop. 10.30], therefore $\mathrm{ind}(D)$ is even. It follows then from (12.7) that

$$v_2([\mathbf{r}, \mathbf{a}]) = v_2\big(\exp(B)\big) - v_2\big(\mathrm{ind}(D)\big) \leq v_2\big(\exp(B)\big) - 1 = v_2(m),$$

i.e., we have equality in (12.8). By Lemma 12.4, n is a 2-power and hence $\exp(B) = n$. Conversely, if n is a 2-power and $\exp(B) = n$, then D has index 2 by [KMRT 98, Prop. 10.30]. It follows from Lemma 12.4 that for the character $(\mathbf{r}, \mathbf{a}) = (m, m; 1, 0)$ we have

$$v_2\big(\mathrm{ind}(D) \cdot [\mathbf{r}, \mathbf{a}])\big) = v_2([\mathbf{r}, \mathbf{a}]) + 1 = v_2(m) + 1 = v_2(n) = v_2\big(\exp(B)\big),$$

i.e., (12.7) holds.

Case 2: $v_2\big(\exp(B)\big) = v_2(m)$.

Comparing (12.7) and (12.8), we deduce that D splits and the equality in (12.8) holds. Hence again by Lemma 12.4, n is a 2-power and so $\exp(B) = m = \frac{n}{2}$. Conversely, if n is a 2-power, $\exp(B) = \frac{n}{2}$ and D is split, then for the same character $(\mathbf{r}, \mathbf{a})$ as in Case 1, by Lemma 12.4,

$$v_2\big(\mathrm{ind}(D) \cdot [\mathbf{r}, \mathbf{a}]\big) = v_2([\mathbf{r}, \mathbf{a}]) = v_2(m) = v_2\big(\exp(B)\big),$$

i.e., (12.7) holds. □

REMARK 12.9. Inspection of the proof shows that the only source of reduction of the value of n_G from $2\exp(B)$ to $\exp(B)$ is the divisibility property 10.1.1 for the canonical representation $\rho : G \to \mathbf{SL}_1(D)$.

13. Dynkin index for groups of type B

Let G be a simply connected group of type B_n with $n \geq 1$, i.e., $G = \mathbf{Spin}(q)$ for a nondegenerate quadratic form q of dimension $2n+1$ [KMRT 98, 26.12].

Suppose that $n \geq 2$. The character group T^*_{sep} can be identified with the sublattice $\mathbb{Z}^n + \mathbb{Z}\varepsilon$ of $\mathbb{Q}^n$, where

$$\varepsilon = \frac{e_1 + e_2 + \cdots + e_n}{2}$$

and the e_i are the standard basis vectors of $\mathbb{Z}^n$. The cocharacter group $T_{\text{sep}*}$ is identified with the subgroup in $\mathbb{Z}^n$ of the elements whose coordinates have even sum. A short coroot is $\eta_G = e_1 - e_2$. The Weyl group is a semidirect product of $H = (\mathbb{Z}/2\mathbb{Z})^n$ and the symmetric group S_n: the elements of H change signs of the coordinates and S_n permutes the e_i.

We choose the set of simple roots $e_1 - e_2,\ e_2 - e_3,\ \cdots,\ e_{n-1} - e_n,\ e_n$. The corresponding cone of dominant characters Λ^+ consists of all characters $\sum x_i e_i$ such that $x_1 \geq x_2 \geq \cdots \geq x_{n-1} \geq x_n \geq 0$.

Let C be the center of G. The group C^* consists of 2 elements: 0, λ. The corresponding Tits algebra A_λ is similar to the even Clifford algebra $C_0(q)$ [KMRT 98, §27.B]. The restriction $\bar{\chi}$ of a character $\chi = \sum x_i e_i$ to C satisfies

$$\bar{\chi} = \begin{cases} 0 & \text{if all the } x_i \text{ are integers,} \\ \lambda & \text{if all the } x_i \text{ are semi-integers.} \end{cases}$$

Write n_0 and n_1 for the gcd $N_G\big(W(\chi)\big)$ for all dominant characters χ which restrict to 0 and λ respectively. By Corollary 10.12,

$$n_G = \gcd\big(n_0, n_1 \cdot \operatorname{ind} C_0(q)\big). \tag{13.1}$$

If $\chi = ae_1$, the W-orbit of χ consists of the characters $\pm ae_i$ and we have $N_G\big(W(\chi)\big) = 2a^2$, hence $n_0 \leq 2$.

LEMMA 13.2. $n_0 = 2$.

PROOF. We claim that $N_G\big(W(\chi)\big)$ is even for every character χ with at least two nonzero integer coordinates. The orbit $W\chi$ is the union of 4-element sets $\{\pm \sum x_i e_i, \pm \sum x_i' e_i\}$ such that $x_i = x_i'$ for all i but one, say j, and $x_j = -x_j' \neq 0$. The value N_G of the sum of characters in each set is even. □

LEMMA 13.3. $n_1 = 2^{n-2}$.

PROOF. Clearly, $N_G\big(W(\varepsilon)\big) = 2^{n-2}$. We claim that $N_G\big(W(\chi)\big)$ is divisible by 2^{n-2} for every character χ with semi-integer components. It is sufficient to show that $N_G\big(H(\chi)\big)$ is divisible by 2^{n-2}. If $\chi = ae_1 + be_2 + \ldots$, then

$$N_G\big(H(\chi)\big) = 2^n(a^2 + b^2) \in 2^{n-2}\mathbb{Z}.$$ □

THEOREM 13.4. *Let G be a simply connected group of type B_n with $n \geq 1$, i.e., $G = \mathbf{Spin}(q)$ for a nondegenerate quadratic form q of dimension $2n+1$. Then*

$$n_G = \begin{cases} 1 & \textit{if } n \leq 2 \textit{ and } C_0(q) \textit{ is split,} \\ 2 & \textit{otherwise.} \end{cases}$$

PROOF. If $n \geq 2$, the statement follows from (13.1) and Lemmas 13.2 and 13.3. If $n = 1$, the group G is isomorphic to $\mathbf{SL}_1\big(C_0(q)\big)$ where $C_0(q)$ is a centrla simply algebra of degree 2 [KMRT 98, 26.B] and we can use Theorem 11.5. □

REMARK 13.5. In the case $n \leq 2$, the even Clifford algebra $C_0(q)$ is split if and only if q has (maximal) Witt index n.

REMARK 13.6. If q is a nondegenerate quadratic form of dimension at least 5, the Rost invariant of $G = \mathbf{Spin}(q)$ coincides with the Arason invariant defined in Example 2.3 (see [KMRT 98, pp. 436, 437]).

14. Dynkin index for groups of type C

Let G be a simply connected group of type C_n with $n \geq 1$, i.e., $G = \mathbf{Sp}(A, \sigma)$ for a central simple algebra A of degree $2n$ with a symplectic involution σ [KMRT 98, Th. 26.14]. Let

$$\rho : \mathbf{Sp}(A, \sigma) \hookrightarrow \mathbf{SL}_1(A)$$

be the natural embedding. Since $n_\rho = 1$ (Example 7.11), it follows from Theorem 11.5 and Lemma 10.1(1) that

$$n_G \mid n_\rho \cdot n_{\mathbf{SL}_1(A)} = \exp(A).$$

In the case A splits we have then $n_G = 1$, and in general, $n_G \leq 2$ since $\exp(A) \leq 2$.

The groups T^*_{sep} and $T_{\mathrm{sep}*}$ can be identified with $\mathbb{Z}^n$. A short coroot is $\eta_G = e_n$. The Weyl group is a semidirect product of $H = (\mathbb{Z}/2\mathbb{Z})^n$ and the symmetric group S_n: the elements of H change signs of the coordinates and S_n permutes the e_i.

We choose the set of simple roots $e_1 - e_2$, $e_2 - e_3$, ... , $e_{n-1} - e_n$, $2e_n$. The corresponding cone of dominant characters Λ^+ consists of all characters $\sum x_i e_i$ such that $x_1 \geq x_2 \geq \cdots \geq x_{n-1} \geq x_n \geq 0$.

Let C be the center of G. The group C^* consists of 2 elements: 0, λ. The corresponding Tits algebra A_λ is similar to A [KMRT 98, §27.B]. The restriction $\bar{\chi}$ of a character $\chi = \sum x_i e_i$ to C satisfies

$$\bar{\chi} = \begin{cases} 0 & \text{if } \sum x_i \text{ is even,} \\ \lambda & \text{otherwise.} \end{cases}$$

Write n_0 and n_1 for the gcd $N_G\big(W(\chi)\big)$ for all dominant characters χ which restrict to 0 and λ respectively. By Corollary 10.12,

$$n_G = \gcd\big(n_0, n_1 \cdot \operatorname{ind}(A)\big). \tag{14.1}$$

If $\chi = e_1$, we have $N_G\big(W(\chi)\big) = 1$, hence $n_1 = 1$.

LEMMA 14.2. *The integer n_0 is even.*

PROOF. We need to prove that the integer $N_G\big(W(\chi)\big)$ is even for every $\chi = \sum x_i e_i$ with even sum $\sum x_i$. It is sufficient to prove that $N_G\big(H(\chi)\big)$ is even. We have $N_G\big(H(\chi)\big) = 2^{r-1} x_n^2$, where r is the number of nonzero x_i's. If $r = 1$, since the sum of the x_i's is even we must have x_n even. □

We have proved:

THEOREM 14.3. *Let G be a simply connected group of type C_n with $n \geq 1$, i.e., $G = \mathbf{Sp}(A, \sigma)$ for a central simple algebra A of degree $2n$ with a symplectic involution σ. Then*

$$n_G = \begin{cases} 1 & \text{if } A \text{ is split,} \\ 2 & \text{otherwise.} \end{cases}$$

The set $H^1(F,G)$ is identified with the factor set of the set of invertible elements in

$$\mathrm{Symd}(A,\sigma) = \{a+\sigma(a),\ a \in A\}$$

modulo the equivalence relation: $a \sim a'$ if $a' = \sigma(x)ax$ for some $x \in A^\times$. The Rost invariant R_G takes the class of an element $a \in \mathrm{Symd}(A,\sigma)$ to the cup product $\{\mathrm{Nrp}(a)\} \cup [A]$, where Nrp is the Pfaffian norm [KMRT 98, pp. 440, 441].

15. Dynkin index for groups of type ^1D or ^2D

Let G be a simply connected group of classical type D_n with $n \geq 4$, i.e., $G = \mathbf{Spin}(A,\sigma,f)$ for a central simple algebra A of degree $2n$ with a quadratic pair (σ,f), cf. [KMRT 98, Th. 26.15] (If $\mathrm{char}\, F \neq 2$, we have simply $G = \mathbf{Spin}(A,\sigma)$, where σ is an orthogonal involution on A.) Consider the composition

$$\rho : \mathbf{Spin}(A,\sigma,f) \xrightarrow{\alpha} \mathbf{O}^+(A,\sigma,f) \hookrightarrow \mathbf{SL}_1(A),$$

where α is the standard isogeny to the special orthogonal group. Since by Example 7.15, $n_\rho = 2$, it follows from Theorem 11.5 and Lemma 10.1(1) that

$$n_G \mid n_\rho \cdot n_{\mathbf{SL}_1(A)} = 2\exp(A).$$

In the case that A splits, this implies that $n_G \mid 2$, and in general, $n_G \mid 4$ since $\exp(A) \mid 2$.

Let Z/F be the discriminant quadratic étale F-algebra $\big(=$ the center of the Clifford algebra $C(A,\sigma,f)\big)$.

The character group T^*_{sep} can be identified with $\mathbb{Z}^n + \mathbb{Z}\varepsilon \subset \mathbb{Q}^n$, where

$$\varepsilon = \frac{e_1 + e_2 + \cdots + e_n}{2}$$

and the e_i are the standard basis vectors of $\mathbb{Z}^n$. The group of cocharacters $T_{\mathrm{sep}*}$ is identified with the subgroup in $\mathbb{Z}^n$ of the elements whose coordinates have even sum. All the coroots $\pm e_i \pm e_j$ have the same length and we can take $\eta_G = e_1 - e_2$. The Weyl group W is a semidirect product of $H = (\mathbb{Z}/2\mathbb{Z})^{n-1}$ and the symmetric group S_n: the elements of H change signs of an even number of coordinates and S_n permutes the e_i.

We choose the set of simple roots $e_1 - e_2$, $e_2 - e_3, \ldots, e_{n-1} - e_n$, $e_{n-1} + e_n$. The corresponding cone of dominant characters Λ^+ consists of all characters $\sum x_i e_i$ such that $x_1 \geq x_2 \geq \cdots \geq x_{n-1} \geq |x_n|$.

Let C be the center of G. The group C^*_{sep} consists of 4 elements: 0, λ, λ^+ and λ^-, where λ is trivial on the kernel of the isogeny α. The corresponding Tits algebra A_λ is similar to A [KMRT 98, §27.B]. The restriction $\bar\chi$ of a character $\chi = \sum x_i e_i$ to C_{sep} satisfies

$$\bar\chi = \begin{cases} 0 & \text{if all the } x_i \text{ are integers and } \sum x_i \text{ is even,} \\ \lambda & \text{if all the } x_i \text{ are integers and } \sum x_i \text{ is odd,} \\ \lambda^+ \text{ or } \lambda^- & \text{if all the } x_i \text{ are semi-integers.} \end{cases}$$

Groups of type ^1D_n

Suppose that Z splits. Then $C(A,\sigma,f) = C^+ \times C^-$, where C^+ and C^- are central simple algebras over F which are Tits algebras of λ^+ and λ^-. Write n_0, n_1,

n^+ and n^- for the gcd $N_G\big(W(\chi)\big)$ for all dominant characters χ which restrict to 0, λ, λ^+ and λ^- respectively. By Corollary 10.12,

$$n_G = \gcd\big(n_0, n_1 \cdot \mathrm{ind}(A), n^+ \cdot \mathrm{ind}(C^+), n^- \cdot \mathrm{ind}(C^-)\big). \tag{15.1}$$

Consider a dominant character $\chi = \sum x_i e_i$ with integer coordinates. Assume first that only one of the coordinates is nonzero, i.e., $\chi = ae_1$ with $a > 0$. The W-orbit of χ consists of the characters $\pm ae_i$, hence $N_G\big(W(\chi)\big) = 2a^2$. In particular, $n_1 \mid 2$.

Assume now that χ has at least two nonzero coordinates, i.e., $\chi = ae_1+be_2+\cdots$ with $a \geq b > 0$. We claim that $N_G\big(W(\chi)\big)$ is divisible by 4. Consider the subgroup $W' \subset W$ which is the semidirect product of H and the group S_2 which acts by interchanging the first two coordinates. It suffices to show that $N_G\big(W'(\chi')\big)$ is divisible by 4 for every $\chi' \in W\chi$. Each orbit $W'\chi'$ is the union of the following sets of characters:

(1) $\pm ce_1 \pm de_2 + \cdots$ and $\pm de_1 \pm ce_2 + \cdots$ for $c \neq d$ and c, d nonzero;
(2) $\pm ce_1 \pm ce_2 + \cdots$;
(3) $\pm ce_1 \pm de_i + \cdots$ and $\pm ce_2 \pm de_i + \cdots$ for nonzero c, d and $i > 2$.

The value N_G of the sum of characters in these sets is equal to $4(c^2 + d^2)$, $4c^2$ and $4c^2$ respectively and hence is divisible by 4. We have proved:

LEMMA 15.2. *The integer n_0 is divisible by 4 and $n_1 = 2$.*

Now consider the integers n^+ and n^-. All the coordinates of a character χ which restricts to λ^+ or λ^- are semi-integers and in particular are nonzero.

LEMMA 15.3. $n^+ = n^- = 2^{n-3}$.

PROOF. Clearly, $N_G\big(W(\varepsilon)\big) = 2^{n-3}$. We claim that $N_G\big(W(\chi)\big)$ is divisible by 2^{n-3} for every character χ with semi-integer coordinates. It suffices to show that $N_G\big(W'(\chi)\big)$ is divisible by 2^{n-3} for every character χ with semi-integer components. If $\chi = ae_1 + be_2 + \cdots$, then

$$N_G\big(W'(\chi)\big) = 2^{n-1}(a^2 + b^2) \in 2^{n-3}\mathbb{Z}. \qquad \square$$

Lemmas 15.2 and 15.3 and (15.1) then yield the following theorem:

THEOREM 15.4. *Let G be a simply connected group of classical type D_n with $n \geq 4$, i.e., $G = \mathbf{Spin}(A, \sigma, f)$ for a central simple algebra A of degree $2n$ with a quadratic pair (σ, f). If $\mathrm{disc}(\sigma, f)$ is trivial,*

$$n_G = \begin{cases} 2 & \text{if } A \text{ is split;} \\ 2 & \text{if } n = 4 \text{ and at least one of the algebras } A,\ C^+, \text{ and } C^- \text{ is split;} \\ 4 & \text{otherwise.} \end{cases}$$

REMARK 15.5. If $n = 3$ we have $D_3 = A_3$ and the group $G = \mathbf{Spin}(A, \sigma, f)$ of inner type is isomorphic to $\mathbf{SL}_1(C^+)$ where C^+ has degree 4 [KMRT 98, p. 375]. By Theorem 11.5, $n_G = \exp(C^+)$. In other words, as $C^+ \otimes C^+$ is Brauer equivalent to A [KMRT 98, 9.15], we have

$$n_G = \begin{cases} 1 & \text{if } A \text{ and } C^+ \text{ are split;} \\ 2 & \text{if } A \text{ is split but } C^+ \text{ is not split;} \\ 4 & \text{if } A \text{ is not split.} \end{cases}$$

Groups of type ^2D_n

The Galois group $\mathrm{Gal}(F)$ acts on T^*_{sep} through $\mathrm{Gal}(Z/F)$ by the involution

$$\kappa : (x_1, \ldots x_{n-1}, x_n) + \mathbb{Z} \mapsto (x_1, \ldots x_{n-1}, -x_n) + \mathbb{Z}.$$

The group Δ is a semidirect product of $(\mathbb{Z}/2\mathbb{Z})^n$ and S_n.

THEOREM 15.6. *Let G be a simply connected group of classical type D_n with $n \geq 4$, i.e., $G = \mathbf{Spin}(A, \sigma, f)$ for a central simple algebra A of degree $2n$ with a quadratic pair (σ, f). If* $\mathrm{disc}(\sigma, f)$ *is nontrivial,*

$$n_G = \begin{cases} 2 & \text{if } A \text{ is split;} \\ 4 & \text{otherwise.} \end{cases}$$

PROOF. As in the inner case we have $n_G \mid 2\exp(A)$. It suffices to show that for every character $\chi \in \Lambda^+$, the integer

$$N_G\big(\Delta(\chi)\big) \cdot \mathrm{ind}(A_\chi) \tag{15.7}$$

is even if A is split and is divisible by 4 otherwise.

Assume first that only one of the coordinates of χ is nonzero, i.e., $\chi = ae_1$ with positive integer a. The Δ-orbit of χ consists of the characters $\pm ae_i$, hence $N_G\big(\Delta(\chi)\big) = 2a^2$ is even. Note that χ is stable under the involution κ, hence $F(\chi) = F$ and $A_\chi = A^{\otimes a}$ [KMRT 98, §27.A]. Suppose A is not split. If a is odd, then the algebra $A^{\otimes a}$ is also not split, $\mathrm{ind}(A^{\otimes a})$ is even and hence the integer (15.7) is divisible by 4.

If χ has at least two nonzero components, then as in the inner case we see that $N_G\big(\Delta(\chi)\big)$ is also divisible by 4.

Finally assume that all the components of χ are semi-integers. The orbit $\Delta\chi$ is twice as large as in the inner case, hence as in the proof of Lemma 15.3 we see that $N_G\big(\Delta(\chi)\big)$ is divisible by 2^{n-2} and therefore by 4 since $n \geq 4$. □

REMARK 15.8. If $n = 3$, the group $G = \mathbf{Spin}(A, \sigma, f)$ in the outer case is isomorphic to the special unitary group $\mathbf{SU}(C, \tau)$, where $C = C(A, \sigma, f)$ is the Clifford algebra of (A, σ, f) with canonical involution τ and degree 4 [KMRT 98, p. 375]. The discriminant algebra $D(C, \tau)$ is isomorphic to A [KMRT 98, 15.D]. It follows from Theorem 12.6 that

$$n_G = \begin{cases} 2 & \text{if at least one of two algebras } A \text{ and } C \text{ is split;} \\ 4 & \text{otherwise.} \end{cases}$$

16. Dynkin index for exceptional groups

We keep the notation of 10.9. Let G be a simply connected algebraic group over F. For every dominant character χ, write V_χ for the irreducible representation of G with the highest character χ (defined over F_{sep}). For a representation $\rho : G \to \mathbf{SL}(V)$ we set $n(V) = n_\rho$.

Let $\omega_1, \ldots, \omega_n$ be the fundamental dominant characters for G. We write dominant characters multiplicatively, i.e., $\prod \omega_i^{e_i}$ rather than $\sum e_i\omega_i$. The Galois group $\Gamma = \mathrm{Gal}(F)$ acts on the ω_i's by permutations.

PROPOSITION 16.1. *If G is an absolutely simple simply connected algebraic group with trivial Tits algebras, then*

$$n_G = \gcd\left\{|\Gamma\omega| \cdot n(V_\omega),\ \omega \text{ a fundamental dominant character}\right\}.$$

We say that G has *trivial Tits algebras* if $A_{\bar{\rho}}$ is split as an $F_{\bar{\rho}}$-algebra for every $\bar{\rho} \in C^*_{\text{sep}}$. Quasi-split groups have trivial Tits algebras.

PROOF. Since the Tits algebras of G are trivial, by [Ti 71, Th. 3.3], every Γ-invariant representation of G_{sep} is defined over F.

Let g denote the gcd on the right side of the displayed equation in the statement of the proposition. The representation $\oplus_{\mu \in \Gamma\omega} V_\mu$ over F_{sep} is defined over F; we denote it by $V_{\Gamma\omega}$. Proposition 7.9.5a gives:

$$n(V_{\Gamma\omega}) = |\Gamma\omega| \cdot n(V_\omega).$$

So g is a gcd of the Rost multipliers of a finite set of F-representations, and hence n_G divides g.

For a given dominant character λ, we now construct an F-representation X such that g divides $n(X)$ and

$$\text{ch}_G(X) = \Delta(\lambda) + \sum_{\mu < \lambda} c_\mu \, \Delta(\mu) \tag{16.2}$$

for some $c_\mu \in \mathbb{Z}$ (Δ being the semidirect product of W and Γ).

Case I. Suppose λ is Γ-fixed.

Case Ia. Suppose that λ is a product $(\omega_{i_1} \cdots \omega_{i_k})^e$ where $\{\omega_{i_1}, \ldots, \omega_{i_k}\}$ is a Γ-orbit of fundamental dominant characters. Let Y be the representation $\otimes_{j=1}^k V_{\omega_{i_j}}$ defined over F_{sep}. Since $\{\omega_{i_1}, \ldots, \omega_{i_k}\}$ is a Γ-orbit, Y is defined over F.

Moreover, by the derivation formula

$$n(Y) = k \dim(V_{\omega_{i_j}}) n(V_{\omega_{i_j}}).$$

Since g divides $k\, n(V_{\omega_{i_j}}) = n(V_{\Gamma\omega_{i_j}})$, g divides $n(Y)$. By the derivation formula, g divides $n(X)$ for $X = Y^{\otimes e}$.

Case Ib. Now suppose that λ is any Γ-fixed dominant character. Then we may write it as a product $\lambda = \lambda_1 \cdots \lambda_r$ where the λ_i are as in Case Ia for $1 \leq i \leq r$. Set X_i to be the F-representation constructed above for the dominant character λ_i. By the derivation formula, g divides $n(X)$ for $X = \otimes_{i=1}^r X_i$.

Case II. Now suppose that λ is *not* Γ-fixed and $\lambda = \omega_{i_1}^{e_1} \cdots \omega_{i_k}^{e_k}$ such that $\omega_{i_1}, \ldots, \omega_{i_k}$ all lie in the same Γ-orbit.

Set $X_0 = \otimes_{j=1}^k V_{\omega_{i_j}}^{\otimes e_j}$ defined over F_{sep}. Note that since $n(V_{\omega_{i_j}})$ is the same for all j, $n(V_{\omega_{i_j}})$ divides $n(X_0)$ by the derivation formula.

Set X to be the direct sum of the Γ-conjugates of X_0, which is defined over F. The number of Γ-conjugates of X_0 is the same as the size of the orbit $\Gamma\lambda$. This has order 3 or 6 (if G is of type ${}^3D_4$ or ${}^6D_4$) and order 2 otherwise. It is divisible by the size of the orbit $\Gamma\omega_{i_j}$ for all j. Then $n(X)$ is divisible by $|\Gamma\omega_{i_j}|\, n(X_0)$. Since $n(X_0)$ is divisible by $n(V_{\omega_{i_j}})$, $n(X)$ is divisible by $|\Gamma\omega_{i_j}|\, n(V_{\omega_{i_j}})$, hence by g.

Case III (the general case). We may write λ as $\lambda = \lambda_0\lambda_1 \cdots \lambda_r$ where λ_0 is Γ-fixed, each λ_i for $1 \leq i \leq r$ is not Γ-fixed and is a product of fundamental dominant characters in the same Γ-orbit, and λ_i, λ_j are products of fundamental dominant characters from different Γ-orbits for $1 \leq i < j \leq r$. Set X_i to be the representation constructed from λ_i in Case Ib for $i = 0$ and Case II for $1 \leq i \leq r$.

The tensor product $X = \otimes_{i=0}^r X_i$ satisfies (16.2) for some c_μ and has $n(X)$ divisible by g by the derivation formula.

We now show that g divides n_G. By Theorem 10.7, it suffices to prove that g divides $N_G\big(\Delta(\lambda)\big)$ for every dominant character λ. We proceed by induction on the partial ordering $<$ of Λ^+. For λ a dominant character, we suppose that g divides $N_G\big(\Delta(\mu)\big)$ for every dominant character μ such that $\mu < \lambda$. The existence of a representation X as in (16.2) proves that g divides $N_G\big(\Delta(\lambda)\big)$. □

Now it is easy to compute the Dynkin index for G with trivial Tits algebras. It depends only on the Galois action on the Dynkin diagram of G and the indices of the Tits algebras. Thus the Dynkin index of such a G is the same as that of any quasi-split group of the same type, where the quasi-split group may be defined over any field. It only remains to compute Rost multipliers of the fundamental irreducible representations of the split group over $\mathbb{C}$. These can be found in the literature (see [Ha 66] and [LS 97]) or may be computed by the computer program LiE.

16.3. Lower bounds. It follows directly from the definition of the Dynkin index n_G as the order of the Rost invariant that for any extension K of F, n_{G_K} divides n_G. This applies in particular to the case where K is a generic quasi-splitting field for G over F as in [KR 94]. Hence, for G^q the unique quasi-split inner form of G, n_{G^q} always divides n_G.

16.4. Upper bounds. Let K be a finite separable extension of F. We have

$$[K:F]R_G = \operatorname{Cor}_{K/F}\operatorname{Res}_{K/F} R_G = \operatorname{Cor}_{K/F} R_{G_K}.$$

Thus, $[K:F]n_{G_K}R_G = 0$ and n_G must divide $[K:F]n_{G_K}$.

Groups of type ${}^3D_4$ or ${}^6D_4$

Theorem 16.5. *Let G be a simply connected group of type ${}^3D_4$ or ${}^6D_4$. Then*

$$n_G = \begin{cases} 6 & \text{if all Tits algebras of } G \text{ are trivial,} \\ 12 & \text{otherwise.} \end{cases}$$

Proof. Suppose first that all Tits algebras of G are trivial. We can apply Proposition 16.1 and use the table in [LS 97, p. 504]: the Galois group Γ acts transitively on the three end vertices of the Dynkin diagram, each of which corresponds over an algebraic closure to a fundamental representation with Rost multiplier 2. The central vertex has Rost multiplier 12, hence

$$n_G = \gcd\{3\cdot 2, 12\} = 6.$$

Suppose now that G has a nontrivial Tits algebra. Let L be a separable cubic extension of F over which G_L is of type ${}^1D_4$ or ${}^2D_4$, so that $G_L \cong \mathbf{Spin}(A,\sigma,f)$ for A an algebra which is F-isomorphic to the Tits algebra of G, endowed with a quadratic pair (σ, f). If G_L is of type ${}^1D_4$, then furthermore the even Clifford algebra $C_0(A,\sigma,f)$ is isomorphic to a direct sum of two nonsplit central simple algebras of degree 8 (in fact, they have the same index as A), so by Theorem 15.4, $n_{G_L} = 4$. If G_L is of type ${}^2D_4$, then again we have $n_{G_L} = 4$ by Theorem 15.6 since A is nonsplit.

By the upper and lower bounds discussed above, we have that 4 divides n_G which divides 12.

This quasi-split inner form G^q of G has $n_{G^q} = 6$. By 16.3, 6 and hence $\operatorname{lcm}(4,6) = 12$ divides n_G, and we are done with this case. □

Groups of type E_6 or E_7

THEOREM 16.6. *Let G be a simply connected group of type E_6. Then*

$$n_G = \begin{cases} 6 & \text{if } G \text{ is of inner type } {}^1E_6, \\ 12 & \text{otherwise.} \end{cases}$$

PROOF. Assume first that all Tits algebras of G are trivial. If G is of type ${}^1E_6$, then $n_G = n_{G^{\text{split}}} = 6$. If G is of type ${}^2E_6$, the orbits of Γ on the set of fundamental dominant characters are two singletons with Rost multiplier 24 and 1800 and two double orbits, where each element of each orbit has Rost multiplier 6 or 150 (see [LS 97, p. 504]). Hence, by Proposition 16.1,

$$n_G = \gcd\{24, 1800, 2 \cdot 6, 2 \cdot 150\} = 12.$$

Assume now that G has a nontrivial Tits algebra. Let G be of type ${}^1E_6$. The split group of type E_6 has Dynkin index 6, so 6 divides n_G.

The Tits algebras for G have index dividing 27 [Ti 71, 6.4.1], so there is some separable extension K of F which kills the Tits algebras and such that $[K : F]$ divides 27. This gives an upper bound: n_G divides $27n_{G_K} = 27 \cdot 6$.

Another upper bound for n_G is given by the Rost multiplier of the adjoint representation, and we may compute this Rost multiplier for the split group over $\mathbb{C}$. Numbering the fundamental characters as in [Bou 81], the adjoint representation has highest character the maximal root ω_2. The Rost multiplier of this representation is found to be 24. Consequently, n_G divides $\gcd(4 \cdot 6, 27 \cdot 6) = 6$, which completes this case.

Let G be of type ${}^2E_6$. Again, considering the separable quadratic extension over which G becomes of type ${}^1E_6$ gives an upper bound of 12 for n_G. Since the quasi-split group of type ${}^2E_6$ has Dynkin index 12, this is a lower bound as well. □

THEOREM 16.7. *Let G be a simply connected group of type E_7. Then $n_G = 12$.*

PROOF. Here we have the lower bound: 12, the Dynkin index of the split E_7, divides n_G.

The adjoint representation has highest character ω_1, and this representation has Rost multiplier 36.

The Tits algebras for a group of type E_7 have index dividing 8 [Ti 71, 6.5.1]. So there is some separable extension K of F of degree dividing 8 such that all Tits algebras of G_K are trivial. Thus n_G divides $8n_{G_K} = 8 \cdot 12$.

Consequently, n_G divides $\operatorname{lcm}(3 \cdot 12, 8 \cdot 12) = 12$, and we are done. □

Groups of type E_8, F_4, or G_2

THEOREM 16.8. *Let G be a simply connected group of type E_8. Then $n_G = 60$.*

PROOF. The Tits algebras of a group of type E_8 are trivial and the absolute Galois group Γ acts trivially on the Dynkin diagram. Thus the Dynkin index is the same as for the split group over $\mathbb{C}$, hence $n_G = 60$. □

The same argument applies in the proofs of the following two theorems.

THEOREM 16.9. *Let G be a simply connected group of type F_4. Then $n_G = 6$.*

THEOREM 16.10. *Let G be a simply connected group of type G_2. Then $n_G = 2$.*

APPENDIX A

The groups $H^{d+1}(F, \mathbb{Q}/\mathbb{Z}(d))$

The purpose of the appendix is to define groups

$$H^{d+1}(F, \mathbb{Q}/\mathbb{Z}(d)) \quad \text{and} \quad H^{d+1}(F, \mathbb{Z}/n\mathbb{Z}(d))$$

for every field F and all $d \geq 0$, $n \geq 1$. We set

$$H^{d+1}(F, \mathbb{Q}/\mathbb{Z}(d)) = \coprod_{p \text{ prime}} H^{d+1}(F, \mathbb{Q}_p/\mathbb{Z}_p(d)),$$

$$H^{d+1}(F, \mathbb{Z}/n\mathbb{Z}(d)) = \coprod_{p \text{ prime}} H^{d+1}(F, \mathbb{Z}/p^{v_p(n)}\mathbb{Z}(d)),$$

where

(A.1) $$H^{d+1}(F, \mathbb{Q}_p/\mathbb{Z}_p(d)) \quad \text{and} \quad H^{d+1}(F, \mathbb{Z}/p^m\mathbb{Z}(d))$$

are the groups defined below. We fix a prime integer p.

The case where $p \neq \operatorname{char} F$

We define the following Γ-modules, where $\Gamma = \operatorname{Gal}(F)$:

$$\mathbb{Z}/p^m\mathbb{Z}(d) = (\boldsymbol{\mu}_{p^m})^{\otimes d}, \quad \mathbb{Q}_p/\mathbb{Z}_p(d) = \varinjlim_m \mathbb{Z}/p^m\mathbb{Z}(d),$$

where $\boldsymbol{\mu}_{p^m}$ is the group of p^m-th roots of unity in $F_{\mathrm{sep}}^\times$. The groups in (A.1) are the Galois cohomology groups. Clearly,

$$H^{d+1}(F, \mathbb{Q}_p/\mathbb{Z}_p(d)) = \varinjlim_m H^{d+1}(F, \mathbb{Z}/p^m\mathbb{Z}(d)).$$

EXAMPLE A.2. (1) $H^1(F, \mathbb{Q}_p/\mathbb{Z}_p(0)) = \operatorname{Hom}_{\mathrm{cont}}(\Gamma, \mathbb{Q}_p/\mathbb{Z}_p)$;
(2) $H^2(F, \mathbb{Q}_p/\mathbb{Z}_p(1)) = \varinjlim_m H^2(F, \boldsymbol{\mu}_{p^m}) = \operatorname{Br}(F)\{p\}$.

[We identify the Brauer group $\operatorname{Br}(F)$ (= equivalence classes of central simple algebras) with $H^2(F, \mathbf{G}_{\mathrm{m}})$ via the cross product construction as in [KMRT 98, p. 397].]

If $d \leq 2$, $H^{d+1}(F, \mathbb{Z}/p^m\mathbb{Z}(d))$ can be identified with the subgroup of all elements in $H^{d+1}(F, \mathbb{Q}_p/\mathbb{Z}_p(d))$ of order dividing p^m [MS 82].

The connecting homomorphism of Galois cohomology groups associated to the short exact sequence of Γ-modules

$$1 \to \boldsymbol{\mu}_{p^m} \to F_{\mathrm{sep}}^\times \xrightarrow{p^m} F_{\mathrm{sep}}^\times \to 1$$

gives a homomorphism

$$h : F^\times = H^0(F, F_{\mathrm{sep}}^\times) \to H^1(F, \boldsymbol{\mu}_{p^m}).$$

For every $r \geq 0$, the composition

$$(F^\times)^{\otimes r} \xrightarrow{h^{\otimes r}} H^1(F, \boldsymbol{\mu}_{p^m})^{\otimes r} \xrightarrow{\cup} H^r\big(F, \mathbb{Z}/p^m\mathbb{Z}(r)\big)$$

factors through the *norm residue homomorphism* [Ta 76, Th. 3.1]

$$K_r(F) \to H^r\big(F, \mathbb{Z}/p^m\mathbb{Z}(r)\big),$$

where $K_r(F)$ is the *Milnor's K-group* defined as the factor group of $(F^\times)^{\otimes r}$ modulo the subgroup generated by the tensors $a_1 \otimes a_2 \otimes \cdots \otimes a_r$ such that $a_i + a_j = 1$ for some $i \neq j$ [Mi 70]. In particular, $K_0(F) = \mathbb{Z}$ and $K_1(F) = F^\times$. We write $\{a_1, a_2, \ldots, a_r\}$ for the image of the tensor $a_1 \otimes a_2 \otimes \cdots \otimes a_r$ in $K_r(F)$.

The norm residue homomorphism together with the cup product in Galois cohomology induce the cup products

$$K_r(F) \otimes H^{d+1}\big(F, \mathbb{Z}/p^m\mathbb{Z}(d)\big) \to H^{r+d+1}\big(F, \mathbb{Z}/p^m\mathbb{Z}(r+d)\big),$$

$$K_r(F) \otimes H^{d+1}\big(F, \mathbb{Q}_p/\mathbb{Z}_p(d)\big) \to H^{r+d+1}\big(F, \mathbb{Q}_p/\mathbb{Z}_p(r+d)\big).$$

For a field L with a discrete valuation with residue field E such that $\operatorname{char} E \neq p$, there is a residue homomorphism

$$r : H^{d+1}\big(L, \mathbb{Q}_p/\mathbb{Z}_p(d)\big) \to H^d\big(E, \mathbb{Q}_p/\mathbb{Z}_p(d-1)\big),$$

see 7.11 in Serre's part of this book.

The case where $p = \operatorname{char} F$

We set[a]

$$H^{d+1}\big(F, \mathbb{Q}_p/\mathbb{Z}_p(d)\big) = H^2\big(F, K_d(F_{\mathrm{sep}})\big)\{p\}.$$

EXAMPLE A.3. (1) $H^1\big(F, \mathbb{Q}_p/\mathbb{Z}_p(0)\big) = H^2(F, \mathbb{Z})\{p\} = H^1(F, \mathbb{Q}/\mathbb{Z})\{p\} = \operatorname{Hom}_{\mathrm{cont}}\big(\Gamma, \mathbb{Q}_p/\mathbb{Z}_p\big)$;

(2) $H^2\big(F, \mathbb{Q}_p/\mathbb{Z}_p(1)\big) = H^2(F, F_{\mathrm{sep}}^\times)\{p\} = \operatorname{Br}(F)\{p\}$;

(3) If F is a perfect field, the group $K_d(F_{\mathrm{sep}})$ is uniquely p-divisible for $d \geq 1$ and therefore, $H^{d+1}\big(F, \mathbb{Q}_p/\mathbb{Z}_p(d)\big) = 0$ for $d \geq 1$.

For every $m \geq 0$ we write $H^{d+1}\big(F, \mathbb{Z}/p^m\mathbb{Z}(d)\big)$ for the subgroup of all elements in $H^{d+1}\big(F, \mathbb{Q}_p/\mathbb{Z}_p(d)\big)$ of order dividing p^m. By [Iz 91, Th. A] the sequence

$$0 \to K_d(F_{\mathrm{sep}}) \xrightarrow{p^m} K_d(F_{\mathrm{sep}}) \to K_d(F_{\mathrm{sep}})/p^m K_d(F_{\mathrm{sep}}) \to 0$$

is exact. Since $H^1\big(F, K_d(F_{\mathrm{sep}})\big) = 0$ [Iz 91, Lemma 6.6], we have a canonical isomorphism

$$H^{d+1}\big(F, \mathbb{Z}/p^m\mathbb{Z}(d)\big) \simeq H^1\big(F, K_d(F_{\mathrm{sep}})/p^m K_d(F_{\mathrm{sep}})\big). \tag{A.4}$$

The product in $K_*(F_{\mathrm{sep}})$ and the cup product in Galois cohomology induce a cup product

$$K_r(F) \otimes H^{d+1}\big(F, \mathbb{Q}_p/\mathbb{Z}_p(d)\big) \to H^{r+d+1}\big(F, \mathbb{Q}_p/\mathbb{Z}_p(r+d)\big). \tag{A.5}$$

---

[a]This definition is somewhat different from what one finds in the literature. The two definitions agree by the Bloch-Gabber-Kato Theorem [BK 86, Cor. 2.8] and (A.4).

LEMMA A.6. *Let E/F be a Galois field extension with Galois group Δ. Suppose that the p-cohomological dimension of Δ is at most 1. Then the natural homomorphism*

$$H^2(\Delta, K_d(E))\{p\} \to H^2(F, K_d(F_{\mathrm{sep}}))\{p\} = H^{d+1}(F, \mathbb{Q}_p/\mathbb{Z}_p(d))$$

is injective.

PROOF. We embed E into F_{sep} over F. There is an exact sequence [Se 65, Ch. 1,§2.6 b]

$$H^1(E, K_d(E_{\mathrm{sep}}))^{\Delta} \to H^2(\Delta, K_d(F_{\mathrm{sep}})^{\Lambda}) \to H^2(F, K_d(F_{\mathrm{sep}})),$$

where $\Lambda = \mathrm{Gal}(F_{\mathrm{sep}}/E)$. The p-primary component of the first term in the sequence is trivial by [Iz 91, Lemma 6.6]. Using the corestriction homomorphism for Milnor's K-groups one shows that the kernel and cokernel of the natural homomorphism $K_d(E) \to K_d(F_{\mathrm{sep}})^{\Lambda}$ are torsion groups (cf. [Ta 76, Lemma 3.4]). As the p-cohomological dimension of Δ is at most 1, the natural homomorphism

$$H^2(\Delta, K_d(E)) \to H^2(\Delta, K_d(F_{\mathrm{sep}})^{\Lambda})$$

is isomorphism on the p-primary components. □

Let L be a field of characteristic p with a discrete valuation with residue field E. We write $\widehat{L}$ for the completion of L. Let $\widehat{L}_{\mathrm{nr}}$ be an unramified extension of $\widehat{L}$ with residue field E_{sep}, so that the Galois group $\Delta = \mathrm{Gal}(\widehat{L}_{\mathrm{nr}}/\widehat{L})$ is naturally isomorphic to $\mathrm{Gal}(E)$. The latter group is of p-cohomological dimension at most 1 [Se 65, Prop. II.§2.3]. Hence by Lemma A.6, the group $H^2(\Delta, K_d(L_{\mathrm{nr}}))\{p\}$ is identified with a subgroup in $H^{d+1}(\widehat{L}, \mathbb{Q}_p/\mathbb{Z}_p(d))$. We denote it by $H^{d+1}_{\mathrm{nr}}(\widehat{L}, \mathbb{Q}_p/\mathbb{Z}_p(d))$. We write $H^{d+1}_{\mathrm{nr}}(L, \mathbb{Q}_p/\mathbb{Z}_p(d))$ for the inverse image of $H^{d+1}_{\mathrm{nr}}(\widehat{L}, \mathbb{Q}_p/\mathbb{Z}_p(d))$ under the natural homomorphism

$$H^{d+1}(L, \mathbb{Q}_p/\mathbb{Z}_p(d)) \to H^{d+1}(\widehat{L}, \mathbb{Q}_p/\mathbb{Z}_p(d)).$$

The cup product (A.5) restricts to a cup product

$$K_r(L) \otimes H^{d+1}_{\mathrm{nr}}(L, \mathbb{Q}_p/\mathbb{Z}_p(d)) \to H^{r+d+1}_{\mathrm{nr}}(L, \mathbb{Q}_p/\mathbb{Z}_p(r+d)).$$

The residue homomorphism in Milnor's K-theory [Mi 70]

$$K_d(\widehat{L}_{\mathrm{nr}}) \to K_{d-1}(E_{\mathrm{sep}})$$

is Δ-equivariant, therefore it induces the *residue homomorphism* as the composition

$$r : H^{d+1}_{\mathrm{nr}}(L, \mathbb{Q}_p/\mathbb{Z}_p(d)) \to H^{d+1}_{\mathrm{nr}}(\widehat{L}, \mathbb{Q}_p/\mathbb{Z}_p(d)) \to H^d(E, \mathbb{Q}_p/\mathbb{Z}_p(d-1)).$$

Note that the residue homomorphism is not defined on the whole group $H^{d+1}(L, \mathbb{Q}_p/\mathbb{Z}_p(d))$.

Properties

We have defined the functor

$$H^{d+1}(*, \mathbb{Q}/\mathbb{Z}(d)) : \mathbf{Fields}_{/F} \to \mathbf{Abelian\ Groups}.$$

In particular,

$$H^1(F, \mathbb{Q}/\mathbb{Z}(0)) = \mathrm{Hom}_{\mathrm{cont}}(\Gamma, \mathbb{Q}/\mathbb{Z}) \quad \text{and} \quad H^2(F, \mathbb{Q}/\mathbb{Z}(1)) = \mathrm{Br}(F).$$

For a finite separable field extension L/F there is the *corestriction* homomorphism

$$\mathrm{Cor}_{L/F} : H^{d+1}\big(L, \mathbb{Q}/\mathbb{Z}(d)\big) \to H^{d+1}\big(F, \mathbb{Q}/\mathbb{Z}(d)\big).$$

We have the *cup product* homomorphism

$$K_r(F) \otimes H^{d+1}\big(F, \mathbb{Q}/\mathbb{Z}(d)\big) \to H^{r+d+1}\big(F, \mathbb{Q}/\mathbb{Z}(r+d)\big). \tag{A.7}$$

We have also defined the groups $H^{d+1}\big(F, \mathbb{Z}/n\mathbb{Z}(d)\big)$ so that

$$H^{d+1}\big(F, \mathbb{Q}/\mathbb{Z}(d)\big) = \varinjlim_n H^{d+1}\big(F, \mathbb{Z}/n\mathbb{Z}(d)\big).$$

If $d \leq 2$, the group $H^{d+1}\big(F, \mathbb{Z}/n\mathbb{Z}(d)\big)$ is identified with the subgroup of all elements of order dividing n in $H^{d+1}\big(F, \mathbb{Q}/\mathbb{Z}(d)\big)$.

Let L be a field with a discrete valuation with residue field E. For every prime integer p we set:

$$H^{d+1}_{\mathrm{nr}}\big(L, \mathbb{Q}_p/\mathbb{Z}_p(d)\big) = \begin{cases} H^{d+1}\big(L, \mathbb{Q}_p/\mathbb{Z}_p(d)\big) & \text{if } p \neq \operatorname{char} F, \\ H^{d+1}_{\mathrm{nr}}\big(L, \mathbb{Q}_p/\mathbb{Z}_p(d)\big) & \text{if } p = \operatorname{char} F > 0. \end{cases}$$

We define the group $H^{d+1}_{\mathrm{nr}}\big(L, \mathbb{Q}/\mathbb{Z}(d)\big)$ as the direct sum of $H^{d+1}_{\mathrm{nr}}\big(L, \mathbb{Q}_p/\mathbb{Z}_p(d)\big)$ over all prime p. There is the *residue homomorphism*

$$r : H^{d+1}_{\mathrm{nr}}\big(L, \mathbb{Q}/\mathbb{Z}(d)\big) \to H^{d}\big(E, \mathbb{Q}/\mathbb{Z}(d-1)\big).$$

The cup product (A.7) restricts to a cup product

$$K_r(L) \otimes H^{d+1}_{\mathrm{nr}}\big(L, \mathbb{Q}/\mathbb{Z}(d)\big) \to H^{r+d+1}_{\mathrm{nr}}\big(L, \mathbb{Q}/\mathbb{Z}(r+d)\big).$$

Let F be a subfield of the valuation ring of L. There is a natural field homomorphism $F \to E$. If $a \in H^d\big(F, \mathbb{Q}/\mathbb{Z}(d-1)\big)$, then $a_L \in H^d_{\mathrm{nr}}\big(L, \mathbb{Q}/\mathbb{Z}(d-1)\big)$ and for a prime element $\pi \in L$,

$$r(\{\pi\} \cup a_L) = a_E. \tag{A.8}$$

This formula follows from Rule R3d in [Rost 96] on the p-component if $p \neq \operatorname{char} F$ and directly from the definition of the cup product if $p = \operatorname{char} F > 0$.

The formula (A.8) applied to the field $L = F((t))$, yields:

PROPOSITION A.9. *The natural homomorphism*

$$H^d\big(F, \mathbb{Q}/\mathbb{Z}(d-1)\big) \to H^d\big(F((t)), \mathbb{Q}/\mathbb{Z}(d-1)\big)$$

is injective.

PROPOSITION A.10. *Let X be an irreducible variety over a field F having a regular rational point. Then the natural homomorphism*

$$H^d\big(F, \mathbb{Q}/\mathbb{Z}(d-1)\big) \to H^d\big(F(X), \mathbb{Q}/\mathbb{Z}(d-1)\big)$$

is injective.

PROOF. Let $x \in X$ be a rational regular point. There is a sequence of points x_0, $x_1, \dots, x_n = x$ of X such that x_0 is the generic point of X and x_{i+1} is regular and of codimension 1 in $\overline{\{x_i\}}$ for all $i = 0, 1, \dots, n-1$. Set $F_i = F(x_i)$, so that $F_0 = F(X)$ is the function field of X and $F_n = F$ is the residue field of x. Every field F_i carries a discrete valuation with residue field F_{i+1}. We have residue homomorphisms

$$r_i : H^{d+1}_{\mathrm{nr}}\big(F_i, \mathbb{Q}/\mathbb{Z}(d)\big) \to H^d\big(F_{i+1}, \mathbb{Q}/\mathbb{Z}(d-1)\big).$$

Let π_i be a prime element in F_i. By (A.8),

$$r_i(\{\pi_i\} \cup a_{F_i}) = a_{F_{i+1}}$$

for an element $a \in H^d(F, \mathbb{Q}/\mathbb{Z}(d-1))$ and every $i = 0, 1, \ldots, n-1$. If $a_{F(X)} = a_{F_0} = 0$, then these formulas imply that $a = a_{F_n} = 0$. □

APPENDIX B

Tables of Dynkin indices

This appendix gives the Dynkin index of an absolutely simple simply connected algebraic group G, summarizing the results in Sections 11 through 16.

Dynkin index of classical groups

Type of G	Dynkin index n_G of G	Result	Page
${}^1\!A_n$ $(n \geq 1)$	$G \cong \mathbf{SL}_1(A)$ for a central simple algebra A of degree $n+1$ $n_G = \exp(A)$	11.5	137
${}^2\!A_n$ $(n \geq 2)$	$G \cong \mathbf{SU}(B,\tau)$ for a central simple algebra B of degree $n+1$ and τ unitary $n_G = \begin{cases} \exp(B) & \begin{cases}\text{if } n \text{ is a 2-power and} \\ \exp(B) = n\end{cases} \\ \exp(B) & \begin{cases}\text{if } n \text{ is a 2-power, } \exp(B) = \\ \frac{n}{2}, \text{ and } D \text{ is split}\end{cases} \\ 2\exp(B) & \text{otherwise}\end{cases}$	12.6	142
B_n $(n \geq 1)$	$G \cong \mathbf{Spin}(q)$ for a nondegenerate quadratic form q of dimension $2n+1$ $n_G = \begin{cases} 1 & \text{if } n \leq 2 \text{ and } C_0(q) \text{ is split} \\ 2 & \text{otherwise}\end{cases}$	13.4	143
C_n $(n \geq 1)$	$G \cong \mathbf{Sp}(A,\sigma)$ for a central simple algebra A of degree $2n$ and σ symplectic $n_G = \begin{cases} 1 & \text{if } A \text{ is split} \\ 2 & \text{otherwise}\end{cases}$	14.3	144
${}^1\!D_n$ $(n \geq 4)$	$G \cong \mathbf{Spin}(A,\sigma,f)$ for a central simple algebra A of degree $2n$ with a quadratic pair (σ,f) with trivial discriminant $n_G = \begin{cases} 2 & \text{if } A \text{ is split} \\ 2 & \begin{cases}\text{if } n = 4 \text{ and one of the algebras } A, \\ C^+, \text{ and } C^- \text{ is split}\end{cases} \\ 4 & \text{otherwise}\end{cases}$	15.4	146
${}^2\!D_n$ $(n \geq 4)$	$G \cong \mathbf{Spin}(A,\sigma,f)$ for a central simple algebra A of degree $2n$ with a quadratic pair (σ,f) with non-trivial discriminant $n_G = \begin{cases} 2 & \text{if } A \text{ is split} \\ 4 & \text{otherwise}\end{cases}$	15.6	147

Dynkin index of exceptional groups

Type of G	Dynkin index n_G of G	Result	Page
${}^3D_4$, ${}^6D_4$	$\begin{cases} 6 & \text{if all Tits algebras are trivial} \\ 12 & \text{otherwise} \end{cases}$	16.5	149
${}^1E_6$	6	16.6	150
${}^2E_6$	12	16.6	150
E_7	12	16.7	150
E_8	60	16.8	150
F_4	6	16.9	150
G_2	2	16.10	150

Bibliography

[Ar 75] J. Kr. Arason, *Cohomologische Invarianten quadratischer Formen*, J. Algebra **36** (1975), 448–491.

[Ar 84] ———, *A proof of Merkurjev's theorem*, Canadian Math. Soc. Conf. Proc. **4** (1984), 121–130.

[Ax 65] J. Ax, *Proof of some conjectures on cohomological dimension*, Proc. Amer. Math. Soc. **16** (1965), 1214–1221.

[Bae 01] R. Baeza, *Some algebraic aspects of quadratic forms over fields of characteristic two*, Proceedings of the Conference on Quadratic Forms and Related Topics (Baton Rouge, LA, 2001), extra vol., 2001, pp. 49–63 (electronic).

[Bay 94] E. Bayer-Fluckiger, *Galois cohomology and the trace form*, Jahresber. Deutsch. Math.-Verein. **96** (1994), 35–55.

[Be 81] F.R. Beyl, *The spectral sequence of a group extension*, Bull. Sci. Math. (2) **105** (1981), 417–434.

[BF 91] P. Bayer and G. Frey, *Galois representations of octahedral type and 2-coverings of elliptic curves*, Math. Z. **207** (1991), 395–408.

[BLV 86] P. Bayer, P. Llorente, and N. Vila, $\widetilde{M_{12}}$ *comme groupe de Galois sur* $\mathbf{Q}$, C.R. Acad. Sci. Paris **303** (1986), 277–280.

[BK 86] S. Bloch and K. Kato, *p-adic étale cohomology*, Inst. Hautes Études Sci. Publ. Math., no. 63 (1986), 107–15.

[Bö 90] S. Böge, *Witt-Invariante und ein gewisses Einbettungsproblem*, J. reine angew. Math. **410** (1990), 153–159.

[Bor 55] A. Borel, *Topology of Lie groups and characteristic classes*, Bull. Amer. Math. Soc. **61** (1955), 397–432 (= Oe. 33).

[Bor 91] ———, *Linear Algebraic Groups*, second ed., Graduate Texts in Mathematics, vol. 126, Springer-Verlag, 1991.

[Bou 59] N. Bourbaki, *Algèbre*, Hermann, Paris, 1959.

[Bou 81] ———, *Groupes et Algèbres de Lie*, Masson, Paris, 1981, chap. 4–6.

[Bou 83] ———, *Algèbre Commutative*, Masson, Paris, 1983, chap. 8–9.

[BS 94] E. Bayer-Fluckiger and J-P. Serre, *Torsions quadratiques et bases normales autoduales*, Amer. J. Math. **116** (1994), 1–64 (= Serre, Oe. 163).

[CE 56] H. Cartan and S. Eilenberg, *Homological Algebra*, Princeton Univ. Press, Princeton, N.J., 1956.

[Ch 94] V.I. Chernousov, *A remark on the* (*mod* 5)*-invariant of Serre for groups of type* E_8, Mat. Zametki **56** (1994), pp. 116-121 (Russian). English translation: Math. Notes **56** (1995), pp. 730–733.

[CHK 97] J.-L. Colliot-Thélène, R.T. Hoobler, and B. Kahn, *The Bloch-Ogus-Gabber theorem*, Algebraic K-theory (Toronto, ON, 1996), Fields Inst. Comm., vol. 16, Amer. Math. Soc., Providence, RI, 1997, pp. 31–94.

[Cr 89] T. Crespo, *Explicit constructions of* $\widetilde{A_n}$ *type fields*, J. Algebra **127** (1989), 452–461; erratum, ibid. **157** (1993), 283.

[Cr 90] ———, *Explicit constructions of* $2.S_n$ *Galois extensions*, J. Algebra **129** (1990), 312–319.

[Deli 79] P. Deligne, unpublished notes of IHES lectures, 1979.

[Delz 62] A. Delzant, *Définition des classes de Stiefel-Whitney d'un module quadratique sur un corps de caractéristique différente de* 2, C.R. Acad. Sci. Paris **255** (1962), 1366–1368.

[Dy 52] E. B. Dynkin, *Semisimple subalgebras of semisimple Lie algebras*, Mat. Sbornik N.S. **30(72)** (1952), 349–462. English translation: Amer. Math. Soc. Transl. (2) **6** (1957), 111–244 (= Selected Papers, 175–308).

[EK 94] M. Epkenhans and M. Krüskemper, *On trace forms of étale algebras and field extensions*, Math. Z. **217** (1994), 421–434.

[EKLV 98] H. Esnault, B. Kahn, M. Levine, and E. Viehweg, *The Arason invariant and mod 2 algebraic cycles*, J. Amer. Math. Soc. **11** (1998), 73–118.

[EL 72] R. Elman and T.Y. Lam, *Pfister forms and K-theory of fields*, J. Algebra **23** (1972), 181–213.

[El 82] R. Elman, *On Arason's theory of Galois cohomology*, Comm. Algebra **10** (1982), 1449–1474.

[EM 73] S. Endo and T. Miyata, *Invariants of finite abelian groups*, J. Math. Soc. Japan **25** (1973), 7–26.

[Fa 51] D.K. Faddeev, *Simple algebras over a field of algebraic functions of one variable*, Trudy Mat. Inst. Steklov. **38** (1951), 321–344 (Russian). English translation: Amer. Math. Soc. Transl. (2) **3**, 15–38.

[Fu 84] W. Fulton, *Intersection Theory*, Springer-Verlag, 1984.

[Ga 01] R.S. Garibaldi, *The Rost invariant has trivial kernel for quasi-split groups of low rank*, Comment. Math. Helv. **76** (2001), 684–711.

[GM 99] S.I. Gelfand and Yu.I. Manin, *Homological Algebra*, Springer-Verlag, Berlin, 1999.

[GW 00] R. Goodman and N.R. Wallach, *Representations and Invariants of the Classical Groups*, Encyclopedia of Mathematics and Its Applications, vol. 68, Cambridge Univ. Press, 2000.

[Hä 92] F. Häfner, *Einige orthogonale und symplektische Gruppen als Galoisgruppen über* $\mathbf{Q}$, Math. Ann. **292** (1992), 587–618.

[Ha 66] B. Harris, *Torsion in Lie groups and related spaces*, Topology **5** (1966), 347–354.

[HS 53] G.P. Hochschild and J-P. Serre, *Cohomology of group extensions*, Trans. Amer. Math. Soc. **74** (1953), 110–134 (= Serre, Oe. 15).

[Hu 75] J.E. Humphreys, *Linear Algebraic Groups*, Graduate Texts in Mathematics, vol. 21, Springer-Verlag, 1975.

[Iz 91] O. Izhboldin, *On p-torsion in K_*^M for fields of characteristic p*, in Algebraic K-theory, Adv. Sov. math. **4** (1991), 129–144.

[Ja 71] N. Jacobson, *Exceptional Lie Algebras*, Marcel Dekker, 1971.

[Kahn 84] B. Kahn, *Classes de Stiefel-Whitney de formes quadratiques et de représentations galoisiennes réelles*, Invent. Math. **78** (1984), 223–256.

[Kahn 93] ———, *Descente galoisienne et K_2 des corps de nombres*, K-Theory **7** (1993), 55–100.

[Kahn 96] ———, *Applications of weight-two motivic cohomology*, Doc. Math. **1** (1996), 395–416.

[Kahn 99] ———, *Motivic cohomology of smooth geometrically cellular varieties*, Algebraic K-theory (Seattle, WA, 1997), Amer. Math. Soc., Providence, RI, 1999, pp. 149–174.

[Kato 82] K. Kato, *Symmetric bilinear forms, quadratic forms and Milnor K-theory in characteristic two*, Invent. Math. **66** (1982), 493–510.

[KMRT 98] M.-A. Knus, A.S. Merkurjev, M. Rost, and J.-P. Tignol, *The Book of Involutions*, Colloquium Publications, vol. 44, American Mathematical Society, Providence, RI, 1998, with a preface by J. Tits.

[Kne 69] M. Kneser, *Lectures on Galois cohomology of classical groups*, with an appendix by T. A. Springer, Tata Institute of Fundamental Research, Lectures on Mathematics, No. 47, Bombay, 1969. Notes by P. Jothilingam.

[Knus 91] M.-A. Knus, *Quadratic and Hermitian Forms over Rings*, Springer-Verlag, 1991.

[KR 94] I. Kersten and U. Rehmann, *Generic splitting of reductive groups*, Tohoku Math. J. (2) **46** (1994), 35–70.

[Lam 73] T.Y. Lam, *The Algebraic Theory of Quadratic Forms*, W. A. Benjamin, Inc., Reading, Mass., 1973, Mathematics Lecture Note Series.

[Lang 96] S. Lang, *Topics in Cohomology of Groups*, Lecture Notes in Mathematics, vol. 1625, Springer-Verlag, 1996.

[Le 74] H.W. Lenstra, Jr., *Rational functions invariant under a finite abelian group*, Invent. Math. **25** (1974), 299–325.

[Li 87] S. Lichtenbaum, *The construction of weight-two arithmetic cohomology*, Invent. Math. **88** (1987), 183–215.

[Li 90] ———, *New results on weight-two motivic cohomology*, The Grothendieck Festschrift, vol. III, Birkhäuser Boston, Boston, MA, 1990, pp. 35–55.

[Lo 01] H.Y. Loke, *Lie algebras with S_4 structures*, preprint, 2001.

[LS 97] Y. Laszlo and C. Sorger, *The line bundles on the moduli of parabolic G-bundles over curves and their sections*, Ann. Sci. École Norm. Sup. (4) **30** (1997), 499–525.

[Mer 81] A. S. Merkurjev, *On the norm residue symbol of degree* 2, Dokl. Akad. Nauk SSSR **261** (1981), 542–547.

[Mer 97] ———, *Comparison of the equivariant and the standard K-theory of algebraic varieties*, Algebra i Analiz **9** (1997), 175–214.

[Mes 90] J.-F. Mestre, *Extensions régulières de $\mathbf{Q}(T)$ de groupe de Galois $\widetilde{A_n}$*, J. Algebra **131** (1990), 483–495.

[Mes 95] ———, *Constructions polynomiales et théorie de Galois*, Proc. ICM (Zürich, 1994), vol. I, 318–323, Birkäuser, Basel, 1995.

[Mes 98] ———, *Extensions $\mathbf{Q}$-régulières de $\mathbf{Q}(t)$ de groupe de Galois $6.A_6$ et $6.A_7$*, Israel J. Math. **107** (1998), 333–341.

[MH 73] J. Milnor and D. Husemoller, *Symmetric Bilinear Forms*, Ergebnisse der Mathematik und ihrer Grenzgebiete, vol. 73, Springer-Verlag, New York, 1973.

[Mi 70] J. Milnor, *Algebraic K-theory and quadratic forms*, Invent. Math. **9** (1969/1970), 318–344.

[Milne 80] J.S. Milne, *Étale Cohomology*, Princeton University Press, Princeton, NJ, 1980.

[MS 82] A.S. Merkurjev and A.A. Suslin, *K-cohomology of Severi-Brauer varieties and the norm residue homomorphism*, Izv. Akad. Nauk SSSR Ser. Mat. **46** (1982), 1011–1046, 1135–1136.

[MS 90] ——— and ———, *The group K_3 for a field*, Izv. Akad. Nauk SSSR Ser. Mat. **54** (1990), 522–545.

[Ne 99] J. Neukirch, *Algebraic Number Theory*, Springer-Verlag, 1999.

[NSW 00] J. Neukirch, A. Schmidt, and K. Wingberg, *Cohomology of Number Fields*, Grundlehren der math. Wiss., vol. 323, Springer-Verlag, 2000.

[Pa 98] I.A. Panin, *Splitting principle and K-theory of simply connected semisimple algebraic groups*, Algebra i Analiz **10** (1998), 88–131.

[Pf 66] A. Pfister, *Quadratische Formen in beliebigen Körpern*, Invent. Math. **1** (1966), 116–132.

[PR 85] G. Prasad and M.S. Raghunathan, *On the Kneser-Tits problem*, Comment. Math. Helv. **60** (1985), 107–121.

[PR 94] H.P. Petersson and M.L. Racine, *Albert algebras*, Jordan Algebras (Oberwolfach, 1992), pp. 197–207, de Gruyter, Berlin, 1994.

[PR 95] ——— and ———, *On the invariants mod 2 of Albert algebras*, J. Algebra **174** (1995), 1049–1072.

[PR 96] ——— and ———, *An elementary approach to the Serre-Rost invariant of Albert algebras*, Indag. Math. (N.S.) **7** (1996), 343–365.

[Qu 73] D. Quillen, *Higher algebraic K-theory. I*, Algebraic K-theory, I: Higher K-theories (Battelle Memorial Inst., Seattle, Wash., 1972), pp. 85–147, Lecture Notes in Math., vol. 341, Springer, Berlin, 1973.

[Re 00] Z. Reichstein, *On the notion of essential dimension for algebraic groups*, Transform. Groups **5** (2000), 265–304.

[Rose 61] M. Rosenlicht, *Toroidal algebraic groups*, Proc. Amer. Math. Soc. **12** (1961), 984–988.

[Rost 91] M. Rost, *A* (mod 3) *invariant for exceptional Jordan algebras*, C.R. Acad. Sci. Paris **313** (1991), 823–827.

[Rost 96] ———, *Chow groups with coefficients*, Doc. Math. **1** (1996), 319–393 (electronic).

[Rost 99] ———, *A descent property for Pfister forms*, J. Ramanujan Math. Soc. **14** (1999), 55–63.

[Rost 02] ———, *A Pfister form invariant for étale algebras*, preprint, The Ohio State University, Sept. 2002; `http://www.math.ohio-state.edu/~rost`

[Sah 72] C.H. Sah, *Symmetric bilinear forms and quadratic forms*, J. Algebra **20** (1972), 144–160.

[Sal 82] D. Saltman, *Generic Galois extensions and problems in field theory*, Adv. in Math. **43** (1982), 250–283.

[Sal 84] ______, *Noether's problem over an algebraically closed field*, Invent. Math. **77** (1984), 71–84.

[Sal 95] ______, *Brauer groups of invariant fields, geometrically negligible classes, an equivariant Chow group, and unramified* H^3, Proc. Symp. Pure Math. **58.1** (1995), 189–246.

[Sam 90] H. Samelson, *Notes on Lie Algebras*, second ed., Springer-Verlag, 1990.

[San 81] J.-J. Sansuc, *Groupe de Brauer et arithmétique des groupes algébriques linéaires sur un corps de nombres*, J. reine angew. Math. **327** (1981), 12–80.

[Scha 85] W. Scharlau, *Quadratic and Hermitian Forms*, Springer-Verlag, 1985.

[Schur 11] I. Schur, *Über die Darstellung der symmetrischen und der alternierenden Gruppe durch gebrochene lineare Substitutionen*, J. reine angew. Math. **139** (1911), 155–250 (= Ges. Abh. I, 346–441).

[Sco 87] W. R. Scott, *Group Theory*, second ed., Dover Publications Inc., New York, 1987.

[Se 62] J-P. Serre, *Corps Locaux*, Hermann, Paris, 1962; 3rd ed., 1968. English translation: *Local Fields*, GTM 67, Springer-Verlag, 1979.

[Se 65] ______, *Cohomologie Galoisienne*, Lecture Notes in Mathematics, vol. 5, Springer-Verlag, 1965; 5th ed., révisée et complétée, 1994. English translation of the 5th edition: *Galois Cohomology*, Springer-Verlag, 1997.

[Se 71] ______, *Répresentations Linéaires des Groupes Finis*, Hermann, Paris, 1971; 5th ed., 1998. English translation: *Linear Representations of Finite Groups*, GTM 42, Springer-Verlag, 1977.

[Se 78] ______, *Extensions icosaédriques*, Séminaire de Théorie des Nombres de Bordeaux 1979–1980, no. 19 (= Oe.123).

[Se 84] ______, *L'invariant de Witt de la forme* $\mathrm{Tr}(x^2)$, Comment. Math. Helv. **59** (1984), 651–676 (= Oe. 131).

[Se 91] ______, *Résumé des cours de 1990–1991*, Annuaire du Collège de France (1991), 111–121 (= Oe. 153).

[Se 92a] ______, *Résumé des cours de 1991–1992*, Annuaire du Collège de France (1992), 105–113 (= Oe. 156).

[Se 92b] ______, *Topics in Galois Theory*, Research Notes in Mathematics, vol. 1, Jones and Bartlett Publishers, Boston, MA, 1992. Lecture notes prepared by Henri Damon [Henri Darmon]; distributed by AK Peters, Wellesley, Mass.

[Se 92c] ______, letter to M. Rost, Harvard (3 Dec. 1992).

[Se 94] ______, *Résumé des cours de 1993–1994*, Annuaire du Collège de France (1994), 91–98 (= Oe. 165).

[Se 95] ______, *Cohomologie galoisienne: progrès et problèmes*, Astérisque (1995), no. 227, 229–257, Sém. Bourbaki 1993/1994, no. 783 (= Oe. 166).

[SGA 3] M. Demazure and A. Grothendieck, *Schémas en groupes. III: Structure des schémas en groupes réductifs*, Lecture Notes in Mathematics, vol. 153, Springer-Verlag, 1970.

[SGA 6] P. Berthelot, A. Grothendieck, and L. Illusie, *Théorie des intersections et théorème de Riemann-Roch* , Lecture Notes in Mathematics, vol. 225, Springer-Verlag, 1971.

[Su 84] A. A. Suslin, *Algebraic* K*-theory and the norm residue homomorphism*, Current problems in mathematics, Vol. 25, Akad. Nauk SSSR Vsesoyuz. Inst. Nauchn. i Tekhn. Inform., Moscow, 1984, pp. 115–207 (Russian). English translation: J. Soviet Math. **30** (1985), 2556–2611.

[SV 00] T.A. Springer and F.D. Veldkamp, *Octonions, Jordan Algebras and Exceptional Groups*, Springer-Verlag, Berlin, 2000.

[Sw 69] R.G. Swan, *Invariant rational functions and a problem of Steenrod*, Invent. Math. **7** (1969), 148–158.

[Sw 83] ______, *Noether's problem in Galois theory*, Emmy Noether in Bryn Mawr (Bryn Mawr, Pa., 1982) (B. Srinivasan and J. Sally, eds.), Springer-Verlag, 1983, pp. 21–40.

[Ta 76] J. Tate, *Relations between* K_2 *and Galois cohomology*, Invent. Math. **36** (1976), 257–274.

[Th 86] R.W. Thomason, *Comparison of equivariant algebraic and topological* K*-theory*, Duke Math. J. **53** (1986), 795–825.

[Ti 66] J. Tits, *Classification of algebraic semisimple groups*, Algebraic Groups and Discontinuous Subgroups (Boulder, Colo., 1965) (A. Borel and G. D. Mostow, eds.), Proc. Sympos. Pure Math., vol. 9, Amer. Math. Soc., Providence, RI, 1966, pp. 33–62.

[Ti 71] ———, *Représentations linéaires irréductibles d'un groupe réductif sur un corps quelconque*, J. reine angew. Math. **247** (1971), 196–220.

[Ti 90] ———, *Strongly inner anisotropic forms of simple algebraic groups*, J. Algebra **131** (1990), 648–677.

[Vi 85] N. Vila, *On central extensions of A_n as Galois groups over* $\mathbf{Q}$, Arch. Math. (Basel) **44** (1985), 424–437.

[Vi 88] ———, *On stem extensions of S_n as Galois groups over number fields*, J. Algebra **116** (1988), 251–260.

[Vo 70] V.E. Voskresenskiĭ, *On the question of the structure of the subfield of invariants of a cyclic group of automorphisms of the field $Q(x_1, \cdots, x_n)$*, Izv. Akad. Nauk SSSR Ser. Mat. **34** (1970), 366–375 (Russian). English translation: Math. USSR Izv. **4** (1970), 371–380.

[Vo 73] ———, *Fields of invariants for abelian groups*, Uspekhi Mat. Nauk **28** (1973), 77–102 (Russian); English translation: Russian Math. Surveys **28** (1973), 79–105.

[Vo 77] ———, *Algebraic Tori*, Nauka, Moscow, 1977 (Russian); revised and augmented English translation: Translations of Math. Monographs **179**, AMS, 1998.

[Wi 37] E. Witt, *Theorie der quadratischen Formen in beliebigen Körpern*, J. reine angew. Math. **176** (1937), 31–44 (= C.P., 2–15).

Index of notation

Index of terms